TRESPASSING

WAYNE STATE UNIVERSITY PRESS | DETROIT

TRESPASSING
DIRT STORIES & FIELD NOTES
JANET KAUFFMAN

© 2008 by Wayne State University Press, Detroit, Michigan 48201.
All rights reserved. No part of this book may be reproduced without
formal permission. Manufactured in the United States of America.
12 11 10 09 08   5 4 3 2 1

Library of Congress Cataloging-in-Publication Data

Kauffman, Janet.
Trespassing : dirt stories and field notes / Janet Kauffman.
p. cm. — (Made in Michigan writers series)
ISBN-13: 978-0-8143-3374-7 (pbk. : alk. paper)
ISBN-10: 0-8143-3374-5 (pbk. : alk. paper)
I. Title. II. Series.
PS3561.A82T74 2008
813'.54—dc22
2007039291

This book is supported by the Michigan Council for Arts and Cultural Affairs

Grateful acknowledgment is made to the DeRoy Testamentary Foundation for the
support of the Made in Michigan Writers Series.

This book has been printed on 50 percent postconsumer recycled paper and 30 percent
postconsumer recycled cover stock. Compared to nonrecycled paper, recycled paper requires
fewer trees to produce, is more energy efficient, results in fewer greenhouse gas emissions and
hazardous air pollutants, and generates less solid waste and water pollution.

Designed by Savitski Design
Typeset by Maya Rhodes
Composed in Dante MT and Rosewood Std

CONTENTS

PREFACE

The fiction in the first half of this book and the nonfiction in the second half both center on a rural Michigan community in the crosshairs of the intense national debate on food production, especially the shift to industrial animal factories—Confined Animal Feeding Operations (CAFOs). Southern Michigan is where I live, where I work, where I breathe the air and drink the water—along with the CAFO owners, the immigrant laborers, the auto-parts workers, small farmers, township supervisors, all of us. These stories and essays look at the lines drawn, and the lines crossed, in America's new rural landscape of factory farms, mega-meat and milk, mega-manure.

Some of the stories first appeared, several in somewhat different versions, in *The New England Review, Santa Monica Review,* and *Witness*. Of the nonfiction pieces, "The Fantasy of the Clip Art Farm" was published in *Dissent;* "Letting Go: The Virtue of Vacant Ground" first appeared in *A Place of Sense,* edited by Michael Martone (University of Iowa); and "Buried Water" was included in *The Huron River: Voices from the Watershed,* edited by John Knott and Keith Taylor (University of Michigan Press). Thanks to Al Rudisill and the Ypsilanti Historical Society for several of the images in "Buried Water." Sections of "This Stream,

That Stream" were published by the *Prairie Writers Circle* and as
"Down the Drain" in *Fresh Water: Women Writing on the Great
Lakes*, edited by Alison Swan (Michigan State University Press).
A selection from "Skinhead Agriculture" was published by the
Prairie Writers Circle as "Bare Earth Is Scorched Earth."

Many thanks to the Sierra Club for a Community Action Grant
that kick-started our water-monitoring projects near CAFOs in
southern Michigan.

And thanks without end to the devoted local gang of
volunteers working for watershed protection and sustainable
agriculture—Environmentally Concerned Citizens of South
Central Michigan and the Bean/Tiffin Watershed Coalition.

1 | DIRT STORIES

WITH MY HANDS, SWIMMING

The black and white cow bodies shift to green again, and all those legs stick in the air. The big bellies one after another stretch across the middle and puff up like balloons, tents for the crowds inside. Then there's a huge collapse and sighing. The hard knees give a good kick, every hoof in the air.

We're in the country now. Under the dead cows of June—I call them that all the time—the dead cows of June. They're scraped up day after day and hoisted with front-end loaders. Hey diddle diddle. The cat and the fiddle. Some things you can't get away from. Even in rubber boots, you can slip. The vultures keep an eye out. They're all over this mound.

I came to Michigan with my brother. I call him Benny, this side of the border. The grass is always greener. Hey, don't even picture cows in green pastures. No, they walk on concrete, and their knees bulge like your grandmother's, crawling on cobblestones for whatever crimes. In the milking parlor, I touch their knees, so they know a cushion.

The Dutchman I call boss calls me little worm.

I know Mother Goose, that's English to me. They are live-wire pages, every bit of action you could want. Benny hates nonsense because it's not a real story, he says, it's just words. He's a moviegoer. He reads detective novels, with endings at the end that kill the crooks or lock everybody up behind bars. He talks to the TV. A minute before the end of a show, he'll say, That's right! Now wrap it up!

Every day here, Benny gains weight. He will soon be as wide as he is tall. He drives the boss's pick-up to town and brings back brown bags of hamburgers everybody eats cold. The paper-wrapped patties are very small, so he eats half a dozen.

Benny hauls dead cows. He pumps cow slurry to a pit, then to a tanker. Sometimes he drives the tanker to a field and sprays the dirt black.

With the rest of the crew, we live in an old farmhouse. The boss pulls rent from our pay and gets another check from George W. Bush.

I say to the boss, Explain this economy again?

He says, Let me tell you a thing or two.

I say, Skip it. And he walks away in his clear plastic bio-security booties.

So, let me tell you a thing or two about breakfast. We skip it.

Let me tell you about evenings.

It all depends which way the wind blows. Sometimes we sit on the back porch and listen to frogs in the ditch. We smoke if the air smells bad. We turn out the house lights and pull the fuse on the idiot light on the pole. Right away the frogs tune up, as if we've all disappeared.

There are three different frogs—the cheepers, the beepers, and the creepers. We imitate them, our own chorus. Alonzo does the cheep, Benny the beep, me the creep. The other guys just laugh. We've walked to the ditch with a flashlight, but we've never seen one.

Let me tell you about laundry. We don't hang it on a line—we don't need no stinking laundry! Benny says. We go to town to the laundromat. It's just like at home, except there is no food on the premises and no soap smell in the warm air because of the dry cleaning in the back, with its chemistry lab and presses. Benny jokes it's a meth lab, but there's just one old lady clipping tags on mohair sleeves and running the racks. The back of the laundromat whirs like the milking parlor—motors, pumps, hydraulic gates, the cow ear clips, computer collars.

I've worked in factories—piecework, lines, even robot welders. Moving parts is one thing. Animals lined up day and night, though. I look around and often think I'm dreaming. If I straighten up my back and close my eyes, my lung tissues fill with cows' breath. My mind floats, and I'm right there with them in darkness.

I can see through hide. With my hands, swimming, I can move through the layers of muscle and then touch those bad bones. I count the four stomachs of every cow. Two thousand cows, eight thousand stomachs.

In the cow's first stomach, the boss sticks a magnet. What I

mean is, he gives the cows a magnet pill, and it stays in the first
stomach to catch all the pieces of wire, the nails from who knows
where, the shards of steel from corn ground in a rusted bin in
some other country. He says a cow's first stomach is packed with
metal by the time she dies.

I dig with my hands under the dead cows of June, and I pick
them out, the junk-heaps. They're small globes. I line them up on
the windowsill. Benny says it makes him sick, but he doesn't stop
looking.

We carry the laundry home in black plastic bags. I sort my jeans
and shirts from Benny's and fold them in my box in the corner. I
cover the box with a towel and drop some detergent on the towel,
perfume, a couple of times a day.

Some evenings Benny and I talk about hitchhiking to Canada.
But these days, we probably couldn't get across the border. It's a
world of prisons, and anybody can end up in a cage with a row of
other look-alikes, computer clips on you, and no way out.

I drink water. A body has to. But other liquid—I can see where
it comes from and I can't swallow. The vats, pipes, the udders and
disinfected tubes—everything dirtied and then bleached. My eyes
roll backward, and it's like knowing too many languages all at
once. You know you have to say something, but nonsense comes
out. Hey diddle diddle.

Little worm, wipe up.

The cat and the fiddle. The cow jumped over the moon.

I drink only water. But I eat anything. I turn dirt into shit. I eat
it up. Now you say that's extreme, but when you're here on the

inside, dug down into dark, you don't care about cleansers. The little dog laughed to see such sport.

I tell Benny we're already gone. Our bodies are flowing toward water, which blackens and thickens and crosses all the borders. Little worm moves toward a Great Lake. I eat everything with my hands. Food or not, you name it. The dish ran away with the spoon. I swallow metal, and it grows in my gut.

BASKING

Each spring, they picked up again as lovers. By the end of
September, everything was finished. They took it for granted. Like
any perennial, cut it back in the fall, let it lie.

Summers, they went swimming and worked around the yard.
All ease and good humor. They might have been brother and
sister, except for the fact that they made love at night in Deanna's
apartment over the garage. Max said more than once they could
marry. But they never lasted to winter. They joked in mid-August
about winter lovers, how they'd have to be different sorts of
people, wrapped under layers of down vests and flannel. Max said,
"Where would you be?" and he pressed his palms against the flesh
of her waist, and she kissed his shoulder.

After a cookout in the backyard, they sat on lawn chairs with
coffee, and Deanna's father reminded the two they had the same
dark hair and same flat blue eyes. "Lazy eyes," he said.

"Dreamy eyes," said Deanna.

"Baby blues," said Max.

"So, hell, have a blue-eyed baby," Deanna's father said.

Neither Max nor Deanna had brothers or sisters. Their mothers were dead. No one could see why they didn't stick.

"Maybe we're too much alike," Deanna said.

"Lovers *should* be," Max said.

"True," said Deanna. "But summers are short. We don't weather."

"Yeah, one season leads to the next. They fit together," Max said.

"And we fall apart," Deanna said.

Deanna's father worked for Mobil Oil, supervisor at the huge southeast terminal. His house was next door, and the garage where Deanna lived stood just to the side of the house. Oil tank trucks and double-bottomed gas haulers rolled in and out of the terminal all day, every season. Her father gauged the flow in the pipes from the north, from Canada, against the outflow at the terminal. At night, the whole place was floodlit, with the brilliant blue-white glow of mercury lights. The gates with concertina wire on top were locked, to keep kids away from the white storage tanks and away from the rows of winding stairways. Although now the company called it homeland security.

Deanna and Max looked out her bedroom window over the floodlit tanks. Fog intensified the whiteness, and they talked through humid summer nights about the non-stop moonlight, the bright square falling on the bed. The room was lit up enough to see chairs and the shadows of chairs.

Sometimes, into the middle of the night, Deanna sat in her white nightshirt by the window, basking.

On Labor Day, they picnicked at Hayes State Park. The next morning, or a morning soon after, Max dropped off Deanna at Kroger's and they kissed each other on each cheek in the car, and that was farewell. By December, they'd almost forgotten each other, the way they'd almost forgotten the heat of direct sun.

Max worked at a greenhouse, planting, transplanting, hauling flats. He lived in a white trailer on Medina Road, and he appreciated the idea of seasonal work. But he worried sometimes that he and Deanna would end up routine, more predictable than seasons. It already *was* routine, four summers running.

Deanna, for her part, thought she and Max would eventually become lovers as steady as work at Kroger's. She understood inevitabilities. In spite of their habit of breaking up, she suspected that one year, like a disruption in climate, their love would continue deep into fall and, at last, as the weather grew colder, they would walk outside together, flood the backyard, and let it freeze, glassy, clear enough to see the grass through. Deanna was a good skater. She'd flooded the backyard every winter from the time she was a child. Saturdays and Sundays, she skated in large circles, then spun herself around through smaller and smaller circles until, with her arms out-stretched and lifted over her head, she blurred herself to a spindle. "Marvelous, marvelous," her father applauded. Deanna knew that someday, in her figure skates, she would circle the yard, up to the fences, with Max, or a man like Max, with his long-legged grace and his black skates. "Well," she heard herself saying, "it will either end, or continue."

Through fall and winter, when the greenhouse business tapered off, Max hired himself out for harvesting corn, the small fields on the small farms. He still had his father's old John Deere combine, slow and steady. He had plenty of work until Christmas if the snows weren't too deep. Driving the combine, he watched the cornstalks snap and shatter in the gathering chains. At night he had a recurring winter dream of harvest—fields of naked men and women, extremely frail, standing before an immense, indiscernible machine whose noise like tremendous choirs deafened them all. Max would wake with his hands over his ears. He'd get out of bed, switch on the light, and look quickly into the dresser mirror. Behind his reflection was the fake wooden paneling of the trailer wall and to the side of his shoulder the 40-watt bulb of the light, unshaded. He'd stand there a moment, hugging himself.

Max went out with a few women in winter. He took Dolores Antin sledding, and lying on top of her, skimming down a steep hill, he nearly decided to marry her. But when he leaned to one side and toppled her into the snow, she complained about the cold, and she wasn't kidding. That changed his mind. He always changed his mind.

Deanna, too, dated once in a while. She went once with Garland Sonora to Detroit to hear the symphony. "It should go on forever," she told him. He liked her for that, thinking her wistful, but she gave up on him after a week and told her father, "The man has no stamina."

When Deanna and Max got back together each spring, they didn't say much about winter. She mentioned the skating. He told her once about the nightmare. But whatever else happened, the details didn't concern them. That was all another time and place. They switched to lightweight shirts and resumed.

In April, after their fourth winter apart, Max noticed trout lily leaves spreading over the floor of the woods toward the road, and he thought of Deanna. He thought of her walking around next week, checking the yellow flowers, and he said to himself, no more. It didn't make sense. He hadn't given her a thought for months; she had nothing to do with half of his life. He put himself through long winter terms and then came back, like a kid home from college, to Deanna. No more. He wouldn't take it so easy this summer.

It won't be me who calls, he decided.

A day or two later, when the phone rang, it rang on schedule.

"Hello," he said, his shoulder against the wall.

"Max, it's Deanna."

"I know."

"Could you stop by tonight?" Her voice was formal, polite. Max stood up straighter.

"I was thinking," he said, "that this year we should try—a trial summer apart. We could try it. What do you think?"

"I agree," she said, definitely. "But I want to talk to you, if you have time."

"You have a good winter?" he asked.

"It was a wonderful winter. That's why I want to talk to you." Her voice had a different pitch to it, a higher, lighter tone. Maybe she'd traveled. Maybe she'd gone to school.

"There's a question I want to ask you," she said. "Will you come over?"

"If it's important—"

"It is."

"All right. I won't stay, but I'll stop over."

"Thank you," she said, and hung up.

Max hadn't driven southeast of town for several months. Blocks away he saw changes—a pastel haze in the sky, reflections like embers in windowpanes. Instead of the white floodlights of the terminal yards, there was a pink and orange mushrooming overhead, not the honeyed glow of a fire but a phantasmagoric fallout. They'd changed the lights.

The white storage tanks shimmered a glossy pink. Deanna's father's house and the white aluminum siding of the garage emerged panel by panel, rouged and flaring. When Max stopped the car and stepped out, his shirt and white socks switched to pink. He knew if he smiled, his teeth would be pink.

That's why she called, he thought. She's mad at Mobil.

He went to the side garage door and up the stairs. He knocked at the door.

"Come in," Deanna said.

Inside, the living room was darkened, both shades drawn. There was a candle where a lamp used to be. Deanna in a long brown summer dress sat in a chair by the candle. In the flickers of light, her face was lined, deeply shadowed, with the exactness of fine makeup. Her chin jutted out and the candlelight fell in a strip across the slope of her nose.

"Sit over there," she said.

"I know," Max said. "It's the lights. You want me to shoot them out."

She had taped the window shades shut, with wide masking tape, all around.

"When did they do it?" he asked.

"It's an evil sign," Deanna said. She was looking at the wall.

"It's a damned bad sign," Max said. "So that's the trouble?" He'd never heard her use the word *evil*.

Deanna's hands rested in her lap, relaxed, one on the other. Under her hands, Max saw the corner of a book, a Bible, a green leather-grained testament.

Max breathed through his mouth. The ridged muscle of his diaphragm pulled his stomach, and he felt himself leaning forward, buckling.

"I didn't say anything about trouble," she said. "But everything, Max, is changed. Everything's new, you know, and by this time of night, I'm very tired." She half-closed her eyes.

"Deanna, where's your father? Does he know you're sick?"

"I'm well."

"You're not well." He walked across the room and sat on the floor beside her. With a heaviness in his arms that surprised him, he put his hands on hers.

"This isn't trouble, Max." Her hands didn't move. "My father leaves me alone. He doesn't understand devotion."

Max worried about her vocabulary. "Deanna—look," he said. "All right. We agreed it was finished, right? We're good at that, aren't we? That's good. I'm after something—" he said, and he patted her hands. He listened to his voice, slowing.

"I'm tired," Deanna said and she stood. She blew out the candle and walked into the bedroom.

"Listen," Max said, in the dark. "Let's give this up. We're not even talking." He stayed where he was, on the floor.

Inside the bedroom another candle flared and Max heard the settling of her bed.

"I have a question for you, Max," she said, in her new voice that quavered.

"What is it?"

"Come here, Max."

He walked to the bedroom door.

Deanna lay in her bed, in her clothes, with the covers pulled to her waist. Her hands were crossed on the blanket. She must have practiced laying herself out neatly, tucking herself in.

Deanna turned her head and looked at him. Her blue eyes came into focus, then blurred again, the flat color of the iris spreading, thinning out. Without smiling, she turned her face away, straight ahead, toward the window with its shade pulled.

A small cross was cut in the shade, and its pink light fell on the blanket near her feet.

"You've thickened, Max," Deanna said. "You've thickened, it looks to me."

"I'm the same," he said. "About the same."

"No, Max, I see a thickening."

"This is lousy light," he said. He straightened his back. His throat tightened and he felt the hinges of his jaw beneath his ears.

"I have a question," Deanna said. She stared past her blanketed feet to the small cross of pink light.

"All right. I'm here to answer a question."

"Have you read John?"

"What?"

"John. The book of John." She held out her Bible. He could see the black lettering. "Have you read John?"

"No. That's easy to answer. No."

"You must read John," Deanna said. "This is important." Her hands returned to their place, folded. Her face stirred. Her forehead was moist. "When they changed the lights, it was dark here, a few nights. Well, it had nothing to do with the lights. I know that." She laughed, two or three separate laughs, then stopped. "Everything's changed. I found a—steadfastness."

"Steadfastness," Max said.

He couldn't remember why, before this, they had ever had reason to part.

"You know me well," she was saying, evenly, "and that's why I wanted to ask you to read John if you hadn't read him. It's important to me, and I thought I should tell you what was important to me."

Max shifted his weight from one leg to the other. "Listen, what do you want me to say?" He leaned into the doorframe; his body buckled again.

"Say you will read John."

"That's it?"

"Yes, say you will read John."

"I will read John," Max said. He would have to reconsider the power of words.

"Now kiss me good night."

Max walked to the bed, bent down, and they kissed cheeks, the same as they kissed in early Septembers, parting.

Max blew out the candle and left.

He knew what would happen—Deanna would cut more crosses into the shades. Through the summer her voice would become more musical. She'd breathe by candlelight, days and nights, and she would grow lighter. By winter, she would skate effortlessly on fiery ice.

Halfway down the stairs, Max realized he was sweating. His shirt was damp, front and back.

Out the door, in the orange glow, he stopped and squinted. He breathed loud through his mouth. In the driveway, he took off his shirt. His bare arms glowed like a baby's. He looked soft, he saw that, and he pulled his elbows close to his chest. When he got to the car, he lifted his forearms to his lips, left then right, pressed hard, and kissed each one.

TRESPASSING

"This is bad," Cecilia says. "This pisses me off. Get down in there."

When the ice melts, my stepmother's on the road, checking streams, eyeballing Child's Drain. Today it's flowing green with globs.

Child's Drain is usually a dribble. From a car, it's not much more than a notch of a place, tented over with prickly ash. But from the seat of Cecilia's truck, you can see the S-turns in the water, a signature, cursive upslope, and then lined out through a flat field, direct to the big dairy. The drain is running strong, and it's running lime green.

This is the place where Cecilia's friend Angie in her purple scarves used to climb down with a sample bottle, fill it, slosh it around, then spit on the ground and smile, "Foul!" She'd pack the bottle on ice and take it to the lab.

Since Angie's gone to Iowa and the big hog wars, where she

says she actually gets paid, I do the dirty work for Cecilia. But I always pull on latex gloves. I scrub with Purell Instant Hand Sanitizer.

Cecilia stands on the road and writes in a journal. She notes the time of day, the temp, the look of the water—and believe me she makes distinctions. Clear, relatively clear, somewhat clear, cloudy, somewhat cloudy, murky, very murky, stinking, silty, globby, milky, oily, slick, filmy, foamy, iridescent, brown, dark brown, green. And so on and so forth.

She drives us home, parks the truck near the greenhouses, and gets on the phone. She dials up the EPA and tells three different people in three different offices that she knows what in this world is meant to be green and what is not. "I don't have green hair," she says. "Well, my stepdaughter does, but that's decoration."

"It is not decoration," I say.

She goes on nonstop. "What are you going to do about it? Get over here. Take a look."

She skips the state guys, goes right to the feds.

Cecilia is in her power suit—white shirt, black jeans, and black jean jacket. She always wears that black jacket, even on Saturdays when she loads flats of seedlings and her potted trees—bladdernut, fancy wafer ash—into the back of the truck. She wears black leather work gloves. If her jacket gets streaked with clay, and it always does, she pounds her chest and beats the dirt to dust with her gloves. Still, by the end of the day at the Farmer's Market, her jacket's brown, and then she'll take it off and slap it against the truck, again and again like a carpet.

Cecilia walks around the kitchen with the phone. She swings

her arms from her shoulders, like a tall actor crossing a room, not flimsy. She has black hair that a few times I've seen in braids, but most of the time she lets it loose, flung out from her head and over her shoulders. It is careless hair, although Cecilia is not careless.

She is also not mindful. She does not think things through to the bitter end, not like I do. I keep an eye on her logic, which often skips the steps you'd like to see on the way to a conclusion.

Cecilia assumes things. She assumes some people will never learn. Ever. She assumes these people are the big guys. She assumes they are cutting the usual corners. She assumes, as stepmother, she should not be called Mother, and she declines the mother role. Cecilia won't make and won't offer cookies; she never tells me to be polite.

You can see why we get along.

She does not perk coffee for the EPA guys. But she assumes they'll show up, and when they do, she says, "Sit down. Here are the maps. Here's Child's Drain. This is what's happening."

It's good we get along because my father's dead and Cecilia's all I've got. Well, at home that's all I've got. She lets me run loose. That's what I'd call it—I can walk where I want, I can drive. So I know a few people. And I know more about Child's Drain than bacteria counts, that's for sure. It's good to have a hoard of knowledge, Cecilia would agree. She has files, documents, bins, tubs with manure management plans, aerial photos, printouts of Web pages from the Netherlands, from Iowa and the hog wars, the chicken wars, the other cow wars in Texas and Idaho. She's got topographical maps, maps of every drain in the whole county.

But Child's Drain, that's my baby. Cecilia would not want to know.

★ ★ ★

Child's is a drain—around here that's what they call a stream. Sometimes when it's clear, with bluish pebbles all polished, with those long strips of rippled sand that the flow carves, it looks like good water melting straight from a glacier.

Child's Drain is half way between here and there. Here is home, the greenhouses. There is the biggest dairy, three hangar-barns full of cows nobody sees, the manure pits out back called lagoons that Cecilia will not call lagoons. She will not say manure either. Feces and urine, Cecilia says. Cow feces and urine.

It's all words, that's what I think. I try to keep my mouth shut. There is so much talk about shit, and insults and flipping people off, the newspapers saying manure spill—like a little glass of milk.

What I feel sometimes in my mouth are syllables strange as teeth, unpronounceable words. A language that doesn't include simple phrases, *yard of cloth* or *pat on the back,* but a language that grinds like gravel, real gravel, *grrrzzzzz, brrrzzzkkkk, tzccchhhh,* pages and pages of scarred words, language that swings things around to ground level, the streambed, and into the dirt, far down, wherever water springs from.

★ ★ ★

He didn't know what I was talking about, but I said something about *loss for words* to Paulie, the guy from the dairy who caught me sitting at a bend in Child's Drain last summer. I was sitting far back from the road, out of the right-of-way and in the field, trespassing, just lounging and looking at things. I didn't feel like talking. At first he thought I was from the other big dairy and started right off, smart-mouthed, when I wouldn't tell him my name, "You don't have one of those strange cow names, do you? Bessie? Marabelle the bell cow. Honey. Clumpen. Hoofenwoof?" He wouldn't quit. "Elsie. Schmaltz?"

"You don't name the goddamned cows," I said. "You've got thousands."

He sat down, thinking I knew him, I guess. "Cool hair," he said. "It's not fake. It's a true color in nature."

"Algae," he said. And he nodded. It wasn't an insult.

So meeting at a ditch in the middle of nowhere got us going on names and words and nature. Now that's a bond.

Maybe he made up his name Paulie for all I know. I didn't check him out. I call him Paulie, and he calls me Cool, for Cool Hair, or just C.H.

We never talk in public—that would be crossing the line. There isn't much public anyway. Cecilia doesn't know he exists; she would never trespass. And I bet Paulie tells his crew when he gets back, he was checking the algae in Child's Drain.

* * *

The bend in Child's Drain, upstream from the road, cuts deep into the field. The banks are steep, overgrown with sumac and tangles of multiflora rose. Even the big machinery working the place can't get to this stretch. We have it all to ourselves.

Paulie's a trespasser here, too, anybody would be. You could say we're all trespassing nature, making fake lines, crossing them, ripping into dirt that's not anybody's.

Paulie and I talk and talk. We aren't lovers. We just like to be in the same place at the same time. Like kids, except that we don't make up games. We're more like old people, very old people, right on the brink, but able to get around, and taking their last looks.

"We're out of it," Paulie says. He slides down the side of the ditch. "Out of the loop, gone from the scene."

Child's Drain was dug out, ditched, to drain the fields. It's overworked, not a pretty place, don't think of it that way. There are the blue pebbles sometimes, and in late summer, tall joe-pye-weed keels out over the water. But usually it's muddy or the water stinks, or ice forms with a brownish crust at the edges.

In summer, you slide down under the vegetation, you're in the dark. I tell Paulie it's like going to a club. No dancing, no drugs maybe, but that other side of things where you can't fool yourself with sunlight and scenery.

"Chains and leather," he says.

"Cow cuffs. The country club," I say.

Paulie does not protest. He leans back, slouches. He's one of those guys who looks blurry whatever he wears—gray sweatshirts, brown cargo pants. But he talks with real precision and never mumbles. That's why it's easy with him. Words are like cups we hand back and forth, shells, or those fossil constructions, corals with holes in the middle. When the water's clear, he picks them out from the pebbles in Child's Drain.

"We're party animals," Paulie says, "but in the end, when you get to the big picture, it's all calm." We talk about being dead, maybe because at the drain we're already so far gone. He thinks of death as sleep, a bed in the ground, very peaceful. He believes he will sleep like a rock, like a person covered with blankets or country quilts. He said that, _country quilts_. I think of it as explosion or travel-through-chemistry, burning to ash, and Cecilia or somebody scattering me in bits behind the greenhouse or here in the drain. Paulie sounds tired, maybe with money or with manure, with all of it. The big scene. When he's here, he's killing time, as they say.

"I think death is the same as right now. This," I say. "Just going along. Geologically, I mean."

"Well, now you're getting beyond the big picture," Paulie says.

"There is no beyond the big picture."

"All right," he says, "that's the _fossil_ picture then, not the _person_ picture."

"I could not care less about person pictures," I say. "You'll end up right here with me in the drain, bone bits washed downstream. Doesn't that make you happy?"

"Sitting here, undead, makes me happy," he says. He leans back. "I could sit here for days."

"That's what I mean. That's exactly what I mean. Sitting here. Bunches of chemicals, molecules. Same as being dead."

"Okay, C.H. Okay," he says.

We both stretch out, our heels in the water. Paulie picks a twig off a low branch. "Tulip tree. Sapling," he says. He scrapes at the bark and peels it back, then hands the twig across to me. "Sniff."

It smells like a couple of summer things—the flowers of multiflora, perfumy grass, citrus of prickly ash fruits. In a hundred years, Child's Drain could flow through forest, if machinery steered clear. Tulip trees sixty feet high, with booming flowers, and low on the banks, maybe pawpaw trees. I brought along some pawpaw seeds the last time, big as lima beans, brown and shiny, and stuck them in the dirt. Cecilia had saved some in the refrigerator for propagation. I took half of them.

There's not much chance of forest, though. They'll probably bury the drain in a pipe and plow over it. "Don't you think?" I say to Paulie.

"Probably will." But he's brought seeds this time, too, black-eyed Susans, some deadheads he picked at the house, he says. He pulls them out of a pocket, tosses them near the water, and pushes them in the mud with his heel. "But water loves gravity," he says. "It'll flow."

We talk and my eyes feel heavy, a press of warm air like hands on my face. Paulie covers his eyes with one arm. We drowse. The sun is up there, cooking the foliage, and green light filters down in curtains. The water slips by. We are not in the world we live in, we're lost to all that, gone to a steady state, in collusion with vegetation, shrubbery, spruces stuck where they are through the seasons, all day and all night with their witch-hat tops and swooshing demeanor. There's no place else to go.

SNAIL, SNAIL, SHOOT OUT YOUR HORNS

Lou beats her foot on the ground like the percussionist she is, and she shakes a can of spray paint around, in a circuit over her head. That metal ball mixer you never see pings like crazy. She shakes the can so hard her wrist joint cracks. "I'll give it another couple of coats."

She guns brown paint at the maple tree in her front yard, shooting in circles, covering over the huge orange letters on the trunk—C U T.

"You think Eddie will notice?" she asks.

"Oh, Eddie will notice," I say.

"Yeah, but when?"

* * *

Lou doesn't care, in the end, what happens. She's in for the battle, the tug of war, the racket she can make. She says that's all Nature is, a process, and anything a person does is part of it.

We've been sitting outside every day for a week, sort of guarding the maple tree, but mostly just watching everything else get ripped up, and trying to adjust our eyes to the new views. We sit on her porch, or mine across the road, and watch the action. It's murky hot, and when air moves, it hauls along streams of dust from the roadwork.

"It's an eye-brain discrepancy," Lou claims. "You'll get used to it."

She pushes her hand out, into the open space down the road where there used to be tangles, whole overthrows, of ash trees and wild grape.

Then her mouth clamps shut because it's too loud to talk. One machine on the tread of another—three hydrohoes, giant backhoes on tank bodies, and then a bulldozer—roll past the house. More muscle.

Eddie the foreman brings up the rear, and dust plumes over his yellow Road Commission truck. He's in that cloud and doesn't see us. Sometimes he'll wave out the window to Lou and yell, "Hey! Scary!"

That's what you call a relationship along this road.

Gray clay settles down slow like flour on everything. Lou brushes at her legs. She sticks her fingers in her mouth and wipes some tracks on her thighs.

★ ★ ★

Lou keeps a permanent kink in her hair and she dyes it black. She wears sleeveless black T-shirts all the time, and you can't tell one outfit from the other. She doesn't mind if she looks dirty. Her black nail polish is all chipped, like she spends time digging somewhere, a couple of times a day.

From the back, we'd look like sisters, with the same wide, drop-off shoulders. We swing our arms when we walk, and we wear the same kind of work boots with black rubber soles that grip good in dust.

I'm a little cleaner. Lou says I've got a pallor, especially across the forehead, and that's true. The skin is tight there, scrubbed. I hack my hair short in summer and I hack my hair short in winter.

But in the brain, well, there is completely different terrain.

"Paper is flimsy," she tells me. "Write all the letters you want. Get those petitions signed. See what happens. *Nothing.*"

She signs, but she says, "It's a waste. I'm only doing this for you. Not them. Writing on paper, it's like some *wafer!*" She sticks her tongue halfway out, like a Catholic.

"And talk is so much air, you know it," she says. "Puffs. You go to those meetings and talk, and I bet Eddie and all the commissioners just watch those words float around and out the window."

"Well, *you* talk. You talk all the time."

"It's just noise," Lou says. "Bone-rattles. The little bones of the middle ear."

★ ★ ★

Smoke from the burn piles shifts this way, and Lou leans back against the wall of the house. She's got a cooler set out, and she says, "Here you go." She pours more gin than tonic into two glasses. The ice is so clear, I hold the drink up to one eye and look through.

The maple tree veers. Somebody drives by in a yellow Road Commission truck.

"He waved at us," Lou says. "The idiot."

★ ★ ★

Lou is often someplace else. From what I hear, she's a rowdy performer. But I've never seen her anywhere except on this porch, along this road, and in town once in a while, by chance. She stomps around, and you can see she does not understand repose or ease, except as demise. When she's away, I take care of her dog, Cash. We go for months without running into each other, and it seems to suit us, the intermittence.

Cash lies inside, up against the screen door. He licks the screen every now and then, eating dust. The screen is full of good-sized holes where his claws have snagged. Lou's afraid if she lets him out, he'll go after the heavy machinery, and they'll run him over as a work-site accident.

Cash ran under a hay wagon once and caught his back foot. Even though the leg was mangled, and the bone stuck out, he came through it without a limp. Lou says the dog does not learn from experience.

"Do you?" I ask.

"Oh yeah, every time. I learn you just live like there's no tomorrow."

I assume Lou could be dangerous. I imagine she screams in her shows, throws sharpened sticks or maybe shards of glass, or commits whatever offenses a woman has to commit on stage to give an audience a scare. Something they'll take home and add to their collection.

Lou has a collection of handguns and a box of knives including a gung-ho serrated switchblade that must be illegal. She claims she uses these for sound effects. Still, one time when we were arguing about self-defense and about whether or not there was such a thing as justifiable homicide, and of course she thinks anything a person does is justifiable, and I said I'd never pull a trigger, no matter what, she hauled these things out to the porch and showed them off. "They're just instruments," she said then.

★ ★ ★

Lou loves noise, it's what she has faith in. I don't know what else. "Percussionist, racketeer," she says, "it's the same thing." She believes in collision, hide hits metal. Wood. Gourd. The rattle of rattles, sticks, taps, slaps. Syncopation.

"Why don't you just grunt, then, and not talk?" I ask.

"Well, that's what it amounts to."

But I know she doesn't mean that—she just likes the sound of it!—because she'll hang out here all afternoon and talk, and spit in the yard and talk more. She puts together plenty of words that make sense.

"Machine waves hit the bones of your ear," she says. "Some earth-mover wrecking whatever. But by the time you hear it, by the time it gets to your brain, it's done. There's not much you can do with a goddamn piece of paper to stop it."

★ ★ ★

A surge of engine noise rocks up against the house, and the gin rolls in breakers over the ice cubes—something's gearing up. Lou sets both hands flat on the porch. I hear some trees crack.

"You can feel the things hit," Lou says.

She appreciates the concussion, the collapse. Hydrohoes grasp and pull trees down—a grip at the waist of the tree, the trunk cracks. A slap on the mat, they're down. Then the claw of the hoe gouges through sod at the base of the tree, straight through the big roots, and the stump comes out like a tooth.

The machines are working just beyond the rise, and behind all the crushing and treadwork, their back-up beeps, high-pitched, come through to the middle ear, over and over, small punctuation of the same line.

"Ride it out," Lou says. "You're fighting a lost fight. No—not fighting. *Flailing.* Look at the company you keep."

She means Barry, in the brick house west of here, who wrote with a paint brush on paper and tied signs around all his trees: DO NOT KILL. And then roped a huge sign across his driveway: OR ELSE.

A picture of his place made the newspaper, and apparently Eddie called the sheriff, who drove out to offer Barry some cautionary words.

But the machinery's past Barry's house now. The trees are gone and so is what Eddie calls brush—all the wild cranberry, bittersweet on the fenceline.

Barry and his mother, Ardyth, walked down here yesterday and handed me the petition from their end of the road.

"It's after-the-fact, I know," Ardyth said, "but here it is."

"They're ghost trees," Barry said. His cap was on backwards, and he had a plaid shirt on backwards, too, like a weird surgeon. He leaned on one foot and then on the other, he picked up gravel in the driveway and pitched it, and he kept looking down the road, toward the smoke from the burn heap. "That's them," he said.

He stepped off the porch and walked out into the middle of the road.

Ardyth called, "Barry!" but he took off without her, dragging his feet and making slash marks in the dirt.

Ardyth sat down on the porch step. With her long fingers, she pulled her gray hair back into two handfuls and tied a knot at the base of her neck. She's a very tall, swaying woman, with a long torso. Tree-like, I've always thought.

"What we've got," she said, "is a strip-mine truck route!"

"For you-know-who," Lou said.

"What we've *got*," Ardyth said, swinging her arms, "is a guy who says wildlife jumps out of brush and leaps at cars. That's one thing. And next, what we've *also* got is a test site here. These hydrohoes come free from vendors, to test out before they buy one." Ardyth yanked at her hair and tightened the knot. "I talked to that driver, with the purple ski glasses. He says they needed a road with trees."

"They got it," Lou said.

"And look what *we* got," Ardyth said. "A test site, hell! It's slash and burn, and it *looks* like hell, fire included."

"Eddie looks down this road," Lou said, and her voice tapped out the words, "he sees a brighter future."

Ardyth stood up and rolled her shoulders, shaking that off, and started toward the road. At the maple tree, she stopped, stretched her legs apart, and then bent over in a slow yoga stretch, touching her toes. "I'm surprised nobody's punched anybody out," she said.

"What's the point?" Lou said. She looked over at me. "We could try out the sound effects and make some real noise."

Ardyth said, "Good. You make as much noise as you can."

We walked with her down the road, then up over the rise until we could see the machines, four hydrohoes, two on each side of the road, working the trees. Closer to us, a machine with hydraulic arms was picking up branches, scooping whole trunks. It clamped thick steel fingers around the stumps and dumped them onto the burn pile. Sparks flew up like fireworks, and smoke blew toward us, still hot, with bits of burning leaf scrap spiraling through it.

Barry was nowhere in sight.

Ardyth pulled the neck of her T-shirt up over her mouth, waved, and walked toward her house.

Somebody was down there near the fire, tossing gasoline on it. One explosion hit, then another, in fireballs through the branches, and before our hands could get to our ears, a couple of clouds, pure flame, shot straight up, several stories high, like those pillars of fire you hear about. And then smoke broke out, almost leafy, at the tops, flattened out the flames, and the fire came crashing back down to the trees.

★ ★ ★

Lou checks her e-mail every couple of hours, and in midafternoon she brings back a printout, a note from her son in Detroit. "Here's some words for you," she says. "He sent nightmare notes." She sits down on the porch step. "He says this dream was so real, exactly like what we've got."

She folds the paper in half, and then folds in the sides until the page is a paper plane. Winding back her shoulder, she sails it off the porch, over the yard, and it keeps going, out onto the road. "These aliens landed and turned people into zombies. He was here on this porch, and over in the field, he saw this dust cloud, and in a few minutes, he saw all these graders and earth-movers, what-have-you, all operated by these human zombies, coming across your farm. Across everything, at incredible speed. They were laying a huge highway, a mile wide. And he says the only way anybody could fight them off was to show real, like *excessive*, emotion. You could cry, or fake cry, or laugh. Or, the best thing, he says, was to get out some speakers and blast *opera*!"

A hydrohoe works its way over the rise, and we have to wait a minute to talk.

"Zombie pavers," she says.

"Did the fake crying work?" I say.

"What do you think? It was a *nightmare*!"

Lou stomps into the house, and I lean back against the heat of the wall. Through my shirt, I can feel my shoulder blades on the brick, and when I cross my legs, my ankle bones scrape on concrete. There's skin on the bones, but you can see right through it to all the mechanisms of tendons in my hands, the veins in my feet, the gristle.

"I'm worn out," I say.

Somehow, Lou hears. She sticks her head back out the door.

"Well, goddamn, it helps if you'd just look down the road. Way way way down the road. Past a couple of Eddie's brighter futures, and then keep going. Ice ages, cataclysms, just keep on going. Ripped trees, war zones, you name it, Eddie, crazy Barry, everything long gone, pressed right alongside that heavy machinery into *rock*."

She steps outside, holding Cash back with one foot, and she's piling things up in air, with her hands. "Old broken bones," she says, "old words on paper, every goddamned one, disappeared, and so long gone they're not even history. And then in the end— this is real percussion—an explosion so loud nobody hears, and all of everything burned to dust very lovely in deep space."

Like somebody turning off a switch, it's suddenly completely quiet.

"Noon," Lou says. And she slams the screen door. "Want a lunch break, too?"

For a while, there's no traffic. The smoke settles low, almost like a mist, sifting across the hay field. I set my palms on the concrete porch. Nothing. Lou's quiet. No cars, no trucks.

The road's blocked off with the fallen trees and with the fires. Nobody's going anywhere.

In the calm, more than a few things come into focus—chicory stalks with ratcheted leaves, like desert plants at the edge of the yard. The dead limbs of elms that touch and form the letter W. With a little time and this kind of quiet, you could probably find a whole word.

★ ★ ★

Inside the house, Lou opens and closes some doors. It's so still I hear her feet scuff across the floor. There's a rustling, a hush. She brings out a plate with peanut butter sandwiches and apple slices, and she hands me a bag of chips. "When they start back up," she says, "let's make some noise, too."

"Noise?"

"Don't worry. I just mean maybe I'll bring the marimba out here and hammer for a while. Make them think. Eddie doesn't know what noise *is*!"

"That wouldn't make much noise."

"Well, you know, just to balance things out. The *idea* of the noise. Level the percussive field," Lou says. She stuffs her mouth with a handful of chips and crunches loud.

* * *

In the middle of the afternoon, with everything back in high gear, Lou starts hauling pieces of the marimba out of the house and onto the driveway. She carries out two bundles of the rosewood keys, rolled up in felt. She sets out a box of mallets.

"Help me with the legs," she shouts.

And I go inside her house for the first time.

The place is full of dog hair and leaves—strewn, Lou would probably say, strewn with leaves. Dry leaves and dust heaps in the corners, bits of leaf debris all over the floor. Like a step down that road, into the future where houses don't matter.

The shaded rooms open up, a dining room with a table of fans, and through a door, another large room with wall-mounted fans, a dead-end alley where winds have funneled everything airborne.

There are box fans, oscillating fans, with wood-floor circles around them where the rotation has blown leaves clear. It's cool, a space with its own climate, perpetual autumn.

It's easy to breathe inside, with the leaf smell. Leaf noise.

But there aren't any chairs. Not a sofa. Does she sit on leaves?

I can see how it is for Lou—when she's here, she's hardly here. She wants to level everything.

Lou calls me into a side room, a walk-in closet, where she's unpacking pieces of the marimba. She hands me a set of metal legs on rollers, and I carry them outside.

Beyond, in the field, Barry's wandering around. Ardyth says Barry sleeps on the porch now. He refuses to go inside. He'll eat cereal if she sets it out there, but he's standing guard at night and won't leave his post.

We drop the marimba's supports, its metal struts, into the grooves of the leg pieces, and then we unroll the keys, setting them in their slots. The marimba stands there, rests there, like a reassembly of tree and machine.

Lou sees Barry in the field, and she walks out to the end of the driveway, motioning him over. She shows him her paint job on the tree, and he nods. I can't hear what he's saying, but his head bobs up and down a good while. He takes off his cap.

In company with Barry, Lou looks awkward, her legs bent out, and her knees hinged sideways. She points to the marimba, an introduction. Taking his hand, she leads Barry along to the porch. "You want water?" she shouts. "Or Coke?"

"I'll have a tall glass of water," he says.

Lou goes inside, and Cash turns in circles while the door swings shut. Then he stands at the screen and barks at the road.

Barry sits down beside me. We watch the hydrohoes come over the rise, driving along the shoulders of the road. They're aimed for the elms, the sweetgum tree—those won't take a minute—and then next in line, the maple tree in the yard. It's an old tree, canopied, with one crooked elbow branch, low to the ground, and a split in the bark, a lightning strike scarred shut, midway up the trunk. It shades the road and it shades the porch.

Lou's rooms of leaves will get more light, a lot more, when this is done. She'll have to shut the blinds to get the same effect.

Lou comes back with water for Barry, and she hands him a wooden flute, too. It's one of those lightweight souvenirs, stamped with "Cedar Point" on it, in blue paint.

"Here you go," she shouts.

Barry nods, and he turns his back to the road. He takes up the flute, and it looks like he's playing it.

Out near the mailbox, a yellow truck skids to a stop. Eddie walks over to the guy driving the hydrohoe and they consult, I guess, about dropping the tree, how to miss the electric wires when they pull it down. The machine angles around for a different approach, and Eddie steps back.

The hydrohoe hooks its arm on the tree, revs up, full RPMs, and Eddie's hands swing up to his ears. He doesn't look at Lou.

Over at the marimba, she's picking out mallets. She takes four of them, with large cushioned heads—the kind you'd use for a muffled waltz, a slow serenade. Then she picks up two of the hardheaded ones, very hard rubber, and hands them over to me. "Play the concrete," she shouts. "Sing something."

The machine noise fills every bit of air space in the yard, and even in the brain, the easiest melody's drowned out.

I tap the mallets on the floor of the porch. I can't hear them hit, and I can't even feel when they strike, the rumble from the machinery moves everything more.

Eddie steps back a few more steps and leans against the fender of his truck. Then he looks over. And you know exactly what he sees.

Lou with her eyes shut playing some waltz or god knows what tune she can't even hear in her own head. Barry on the wood flute. Me hammering concrete. I think of a few words, some old rhymes. So I shout them out loud. They're words in the air. Hark, hark, the dogs do bark. Beggars are coming to town.

That's the band Eddie's got when everything comes down.

Lou opens her eyes; she watches my mouth move. Maybe she reads lips. Maybe she'll write the words down on paper and scream them on stage at her next performance. Awake, arise, pull out your eyes, and hear the time of day. And when you've done, pull out your tongue, and see what you can say.

Snail, snail, shoot out your horns. Father and mother are dead. Brother and sister are in the back yard, begging for barley bread.

MONITORING
10 SPOT SAMPLES

Site 1: Lime Lake Inlet

Little Leo stays put at Lime Lake. On the porch, in the boat, at
the culvert. When does he move? He's like the moon. You watch
it and never see it take a step. It's in the east, but then it's in the
west. Leo has a moon face, just a smudge here, another there. He
watches over the water, and neighbors say he's telepathic. Cecilia
says it's true—she gets a call, through the air, check the inlet,
check the inlet. She arrives with sample bottles, and you guessed
it, black water. Little Leo's in his chair, smiling. Or no, he's over
there, looking the other way.

Site 2: The Bunker

Nobody thinks that sitting on concrete is a date anymore. But
Carla does, and she stretches out her legs until her anklebones

stick out. Her knees look good, too, and on the concrete her thighs are more cushy than they'd be on a sofa. She gives her legs a little shake. She leans back against the hard wall of the bunker. That's what Manny calls it, the bunker—where the big dairy stacks up a mountain of chopped hay, covered over with black plastic. The bunker is concrete on three sides, with a concrete slab all around. On the back side of the black mountain, the view stretches across a corn field, and the smell of manure isn't usually too bad.

When Manny comes off his shift, he strolls behind the bunker like he's going for a smoke. Around the corner, there's Carla, resting, in her white shorts and denim shirt.

They have conversations on the concrete, it's that kind of love affair. The kind that won't pan out because he works there, and her family runs the place.

Manny's a reader; he talks about the Middle Ages, and how much this is like that, with different costumes. He pulls off his shirt. He rolls it up because it stinks. He'll wear his own skin instead, 100 percent silk for Carla to touch.

Site 3: Down That Road

If you go down that road, honey, and start talking to those animals, it will goddamn just not stop. You talk to Jada the bird, that's it. You talk to the big Suffolk, the Connemara, you talk to Annabella. Don't talk to cows you can't see. Start up with whatever's passing through, you're done for. Moles, voles, the whole pack of coyotes, common egrets, geese, stray cats and dogs.

I used to talk to Red-Wing, you remember, the one that waited for sunflower seeds in the morning. But I left it at that. Not every goddamned insect. Worm. Low-life.

You have the worst habits. Let somebody else get a word in. What's Jada said to you lately besides I love you, I love you. You believe that? Wait till some bug says, Git! That'd mean something. There's good advice.

Site 4: The Cell Tower

Maggie told Arthur the red blinking light was as bad as his eye tick, the winking he did in meetings. It wrecked the scene. The lost dark was the worst, but daytime wasn't much better. In sunlight, the eye winked silver.

I'm on a blankness campaign, she told him. There's too much action.

Her campaign was, block it. The job took about an hour, and 350 bucks. A tree-moving company brought in heavy trucks with four hydraulic knives that sliced through dirt, down six feet. The blades carved out root balls, and boom arms lifted two tall spruce trees straight up. The trucks drove the trees from one side of the back yard to the other. Maggie stood at the kitchen window and signaled OK when the first tree was in the right spot. Arthur sat on the porch and eyeballed the other one, a little left, that's good.

From those two points, the blankness was back.

Arthur stood up. You won't be able to move your chair much, he said. I can see it from here.

Well then, Maggie said, I'll just have to sit in your lap.

Site 5: Lime Creek

Down in the ditch, there's nothing to see but the slime sometimes,
which is algae maybe, or duckweed, or milkhouse gunk. It bugs
the eyeballs. Just look. Water is murky or silted, or write down
chocolatey, Hershey water, it stinks, it doesn't belong to anybody.
Did it rain here? Cattails lean on one side, and buttonbush on the
other side, with balled-up starships for flowers, like some cartoon
plant. Not much else. Tall bellflower, cleaver, chicory. Maybe a
frog. Live snails, then thousands of dead snails. Dead water. Dead
zones. In the shade there is gray clay mud, a few gravel sandbars
with crinoids from some other time and place. Elderberry, and box
elder. Not much else. False dragon-head, plenty of that. Skunk
cabbage in the bog. Climb in, climb out. Climb in.

Site 6: Toad Creek

Does not turn men into princes. Not with the stinking stream
of black like a scarf cut loose from a mourner. It flows over the
ice, water turned upside down, its shadow on top. A toad is a
toad, and not here. What adornments do we have now, what
croakers? And why are the women propped at the edges, like
stalks? They dip their buckets in, without reflection, down at the
stream with buckets, but not for drinking. What a surprise to be
here, a country no one imagined, a peasantry once again, women
with baskets and buckets, Mexicans with the cows, the palest
men in their counting houses, courtyards, and cab-mounted air-
conditioned machinery always earth-moving, they call it earth-

moving, they claim they're getting somewhere, scraping and
hauling, bunker to bunker, pit to pit.

Site 7: By the Manure Pit

Manifest in view at all times. You know the ropes, huh. Hook
the hose, gun the pump, rev the semi, move it out, hit the gears,
just drive, get there, pump it in the dump box, dump it, next guy
pumps it out, don't worry about that, don't make me say it again,
he knows the ropes, pump it, pump it, spray it on, don't get stuck
in that shit, turn around sharp, give it the gun, get back here
pronto, don't make me wait, breathing this shit.

Site 8: Clam Island

It's a rare thing. A place where without dreaming you see through
water to a landscape of beveled stones and strewn sand, pink and
brown gravels burnished with cold, you are looking into the past,
the remnants of darkest meltwaters pushed through fissures and
finally flowing over your bare feet and over the live clams with
their shells opened in a dark line, cave after cave of shells as they
siphon water, feeding. Sometimes you see a clam foot, fleshy
and pushing through gravel, inching the shell along. Mottled
sculpins swim over them, and even the stick-covered caddisfly
larvae, crawling slow, pass them by. If you stand in the water your
arms cool, air draws across your skin, you're one of those water
plants half in, half out of two climates, or more, some mottled

with light through leaves, or layered by scent, by thought, by the twining of vines, there are clouds you can see both behind the trees and within the water. Colors blur and swirl through the boundaries, your body touches everything. It's impossible to think one thought or remember one memory, they are all there, flooding, and when all things come together, don't be surprised if tears spring up and fall from your face back to water. Such a place has no designs, no intentions, but claims you and calms you, the way your own language, every word of it, shapes itself in your body and exits your mouth as air, or spells out through lettering, page after page what you cannot believe or comprehend.

Site 9: Tamarack Road

Rosanna learned the French word for *storm* from Bud. *Orage.* He showed her how to gargle the *r* in the back of her throat, how to keep it there and not let it out. She wakes up thinking, *orage.* The sounds are serene, drawn out, an easy *ah,* the softened *g, zzzh.* *Ohr-ahzzzh.* Her pulse slows when she shapes the word in her mouth, and her back straightens. This is storm as she understands it—the quiet before, the impending.

She says the word to herself off and on all day, *orage,* whenever she thinks about anything past the garden, past the bergamot blooming in the weeds. Beyond that swirls a stench aerosolized and drifting, obliterating wood mint, ground smell, the blue air. On beyond that, on and on across the map, an immense all-digestive body gathers and grows, like Bunyan releasing liquids from orifices and pores, his blue ox Babe always moving, putting one foot in front of another in every direction, stirring up fine particulates, small and large outbreaks, greasing wheels and commandeering the climate, sliding through landslides and cyclones and then—ah, Rosanna thinks, *orage, orage*—they stop, they squat, there's the terrifying calm that makes every living thing hold its breath.

Rosanna never says the word in public. She keeps *orage,* as she tells Bud, under her hat.

But Rosie, Bud says, you never wear a hat! He puts his hand on her head and rolls strands of her hair around his fingers.

Bud learned French twenty years ago from Annette, who stuffed feathers into fabric sacks in the shapes of birds. One of these birds, with tacks implanted as teeth, hangs from the ceiling near his bed and moves in the wind. He tells Rosanna his one regret was not saving for himself all of the birds, to fill the room and shadow the bed in moonlight, the bed a raft where they'd float, flocks aloft and circling.

They spend hours in bed, sometimes eating breakfast or dinner there. One or the other, they pick peas and cook them up with curried potatoes. Or slice the arrowhead cabbages with parsley and make a slaw.

They do the usual cleaning and painting and patching. Pumping gas in the car. But they test poorly on the idea, make your work a way of life. Bud points to the words in a magazine.

A way of life? says Rosanna.

A way? says Bud. It's a life. We're in the midst.

He is not a chronicler of times past, harms, insults in childhood, scrapes, disgusts. Or anguishes, deaths, nightmares, cataclysms now. He lives like a blind man though he isn't, touching everything with his hands to know it, or sniffing the air, even foul morning fogs off the field, until turning his head, finally, he crumples a handful of spicebush leaves to raise the scent, or in winter he scrapes the skin of twigs to know one from the other, and only these things are his conversation, the phantasms of oils, metals, infusions, timbres, textures, he goes on and on about ephemera and says, that's it, Rosie, don't you worry, that's all there is.

Site 10: Kit's Place

Kit and Carmen are the oldest ones on the road. They go everywhere and hear everything. If somebody's dumping on a new field, they know where, they know when. Kit drives for miles, a square mile block, the next mile block, and Carmen takes notes. They know a couple of manure haulers, they eat dipped cones with Danny every day at Dairyland.

Kit tests the stream at his place once a week. He sends the numbers up the line, county to state to federal. He says this is step three. Step one was do it all—he had cows, he did light industrial, he organized, he did the nuclear waste dump picket, he fished, he caught Carmen. Step two was stop everything—his heart failed, he quit smoking, he took vacations, he stopped worrying. Step three, what the hell, nail the bastards.

A GEOGRAPHY, A LIBRARY
(CECILIA SAYS)

For what is Michigan remarkable?
Where are its populations?
Where are the forests?

Harper's School Geography (1888)

Just past Sword Highway, an electric line catches our headlights
and flashes back lightning into the tree. Which releases the foliage,
and that's the beginning.

The downfall. Veins of leaves ratchet to 3-D, hands opening,
the blood pumping, now pooling in vessels. The green of the
leaves drains to gold, then the gold breaks apart and mottles,
camouflage. The leaves roll over, and the flip sides are blue, they
break through darkness to daylit sky. Cloudless. Until the sky
drops its curtains, closing. From the side, a sheer swag of green
rolls across and gathers into leaf lines again, cell by cell. The edges
curl, they are charred, watercolors wash through black smears to

full fire. The flames hold steady, they fan out like leaves, veined and burning on short stems.

This is the bouquet we carry home and set in a jar.

The sandy banks dissolve like so much sugar, in a single day the course of the river is radically changed, and steamboats pass where a few hours before was cultivated land.

Joel Cook, *America, Vol. 3* (1900)

Out back, six different hillsides slide away into bowls—amphitheaters. When books don't help and protests don't help and even watercress doesn't stir the spirit, we walk to one bowl or the other, and we lie flat on our backs there, cushioned—

1) near the lane, the dip that never holds water
2) where the deer scrape the white pines and drop their antlers in March
3) behind the steep hill where the hay wagon tipped
4) the ice-sledding slope, where it levels off
5) between the staghorn sumac and the coreopsis
6) the dropoff next to the wild asparagus

Root-smell and groundswell fosters us and finally cuts the cord. Take it, the blast of sun on eyebrows, lips, and neck. Passion is for the long term and the ephemeral. Nothing in between. Rocks and water, phantom gnats, bird song. Words with vowels like kisses saying *O,* they're sweetest outside, out of sight, from our own lips. *Metamorphosis. Obliteration.*

Three miles away, pivot irrigators spray liquid cowshit on fields. If we stand up, we catch the ammonia kick in the air, the hit of it, smack, the sting.

So for now, lie low. Lie low.

"Eggscoose me, lady!" he began apologetically.

Julie M. Lippmann, *Martha By-the-Day* (1912)

The fat cat with his gavel, always a black suit. There are very large chairs in the room and only the fattest men fill them.

When he stands, he pushes himself up with both arms, but that's only at the end. From the start, through the middle, and on and on, he spins his chair side to side, pushing off with one foot, braking with the other just when his back is turned.

The tall windows frame twelve segments of sky. Blue. Not a cloud. This room is unusual. When you say clean air, you can look at it out there behind the glass. Most of the time, there's no window, no sky or sectioning of sky. In windowless rooms, any call for clean air is so much wafted breath, passing the foreheads and eyebrows like nothing, and settling on the men's shoulders, pressing the wool blends. They never see it.

We'll have to be looking through scrims of particulates and fecal dust, in a room with fans and a prohibitively expensive, taxpayer-subsidized ventilation system, talking through filtering devices on all our masks, I guess, before he stops spinning.

The progress of the work has been notable, though not as rapid as would be possible if the Department had sufficient authority to properly enforce its regulations.

Report of the Chief, Bureau of Animal Industry,

Department of Agriculture (1889)

Some things are very small. A mote, a mite, a molecule. Some things are so small, they get under your skin and you live with them as yourself. Or they flow in your blood and stir roundabout to your heart, to your lungs, and into your fingertips. When you touch your lover, the smallest things make a cushion and warm your belly and ride with your breath into his mouth.

We say we are living, alive. We say we live in the world, and we mean something by that, counting hours and days.

If we could measure time on the smallest scale, we would not exist. We're so slow, we could never get there. And at the far other end, the longest stretch, we're gone from that, too. Long gone. Such peace, when we're out of the picture. We have these suspicions. Inside we're tumbling like specks and lucky to hold onto anything; outside we move so slowly our stench overtakes us, our destructions whirl past, cement and barbed wire cutting our necks and wrists, there's a swirl of blood in the air we believe is the sunrise when we see it, and we take a picture and frame it and hang the blood on all the walls we can touch.

If the track has regularly spaced pairs or groups of prints, the animal is probably in the weasel family. When these animals leap, they touch the ground with both forefeet and then place their hindfeet in nearly the same prints.

Dorcas Miller, *Track Finder* (1981)

Tricky. It's oh so tricky. Who's the culprit?

The *E. coli* count at South Medina Drain is off the charts.

H. the Investigator looks around, counts 15 geese in the field. 1, 2, 3, 4, 5, . . . 6, 7, 8, . . . 9, 10, 11, 12, 13, . . . 14, 15. She makes a note, *approximately 15 geese in the field.* S. the Dairy Guy nods, it's wild animals for sure—geese, deer, you name it—crossing the field. And the muck ground, too, discolors the drain, he says. Deer in the woods. Other wild animals. Tannins, muck ground.

Somebody says somebody saw a black bear on Camden Road this summer. Eight miles away, but they do roam. Maybe coyotes, maybe turkeys. A. the Supervisor mentions raccoons in the water. He sees raccoons head to tail in the water. J., G., and a host of upperlings concur, or don't dispute.

Everybody looks at the water. There it is. Doesn't look great, but what water does these days? *Brownish water,* notes H.

No smoking gun, she says. Forget water samples. No more *E. coli* tests. Don't even think about DNA. Hey, no spinach or lettuce around here. Late in the day anyway. Don't test, you can't tell.

No smoking gun.

Those four thousand cows? Whew, a big number. No ammunition there. Thousands of gallons of liquid manure on the field? No firearm there.

Hey, they're doing their best, H. says, sometimes you just have to take a person's word.

In French, assimilation is mostly regressive, i.e. the second consonant of the pair influences the first. This, of course, operates in two ways.

Robert J. Gregg, *A Student's Manual of French Pronunciation* (1964)

Talk to me, talk to me, tell me, we'll drive to Lake Michigan one of these days and walk on the beach for thirteen miles, no, farther than that, seventeen miles, all day, hum that lullaby, pack a pack, if we hit a river entering the lake we'll swim across, make it a partly cloudy day, we'll walk barefoot without burning flesh step after step, no reason to make it a trial, hot coals like the other day when the road from one stream to the next smoked our tires, you know what it's like to walk on sand in the morning, moon-cast, cool and gray, how the sand gives, tell me about the time you walked the beach at Assateague, all those miles, the trees standing behind the dunes, the bird part of the story, the hermit crabs, what goes around comes around, last words, the dolphins offshore, keep walking, this is the best part, your legs left and right, and the animals nearby.

CARRIED AWAY

Aurora stepped ahead of the girls into the canoe. She was eighty, tanned from road-walking, her white hair in a buzz cut. Her granddaughters stood in the mud, Ginny holding the rope and Laila the packs.

Aurora shifted her feet for balance, got it, and sat on an orange plastic crate, her seat in the middle. She reached out and grabbed a handful of sedge on shore. Hang on, hang on. A canoe, a torpedo, a rocket, a plane, even these days a goddamn car—it was pretty much the same. When she lifted one foot off the ground and set it down in a metal capsule, she knew she was crossing a line. When she shifted that second foot, it was take off.

She held the canoe steady while Ginny and Laila climbed in, tossed in binoculars.

"This is a ratty canoe," Aurora said.

Ginny stepped around her. "Yeah, ain't it great," she said. She leaned forward, like the carving on the prow of a ship. Her brown hair stuck in strips to her neck, slick from the heat. Nothing was tough enough for Ginny, she wanted to sweat even swimming. She called herself GI Gin sometimes, wore a camouflage T-shirt and pants until they stank and Aurora said, Jesus!

Ginny would rip off her shirt, shower, do laundry, and in a couple of hours start from scratch in the same getup.

Laila settled in the stern, picked up a paddle from between her feet. She was skinny, long-legged, and sinewy in her black T-shirt and black shorts, a bandana on her head, pirate-tie. She raised her arms, did a baton-twirl flourish with the paddle.

They were good to go.

Up front, Ginny shoveled back water, and they pushed off through the heat toward the far side of the marsh. They took the long way around, across the inlet, through flooded cedars, dead for decades and falling apart. Most of the branches were tumbled in heaps, jagged and sheared, clotted with debris. Old trunks lay side by side, caught, half drowned, smooth as bolts of gray silk. A few stumps still standing in the water were riddled with holes.

"Looks like men with guns had a heyday," Ginny said.

"Or bombed their brothers," Laila said.

"Men with guns? Bombs?" Aurora said. "Those are nest holes. Woodpeckers."

"Well, the place looks blasted," Ginny said.

Their water path was a maze, slow going. The sun ricocheted overhead in a blue bowl of sky, and the water shot it back. Ginny paddled, slow chops and cuts, with Laila steering until they cleared the trees. Then the canoe slid into open water, and they paralleled a stretch of cattails near shore.

The sun pressed down hard, the water flattened under that iron, and the air stopped going anywhere. Aurora dragged her fingers in the water, but it didn't cool.

Ginny shook her head to stir up a little air, then pulled off her T-shirt and straightened her long neck. "Sail on!" she said.

"Your tiny tits," Aurora noted, "are hardly a promise of buoyancy."

Ginny didn't look back. She jutted her chin out, aiming forward. "So we sink," she said.

★ ★ ★

Out on the marsh with no edge to it—just a blur into sedges and rushes, a mesh of willows in thickets—Aurora shut her eyes. Heat and quiet. No cow stink, no lawns, no mowers, no pole barns, no rows of corn. Water swooshing by, that was it. Air empty and sweet between the hairs on her head.

She put her hand on her head and said, "Touched."

"What?" Ginny said.

"Touched, you know, everything touches you out here," Aurora said.

"*Touched by Too Many Hands*," Laila said. "A page turner."

★ ★ ★

They're bookworms, Aurora told her friends when the girls were small. Those girls need real action.

But she was a bookworm, too. As the girls grew and read more, the three of them talked books and read out loud to each other in the summer. They sat for hours on the porch until Aurora said, Enough! To the woods!

But even out walking, they collected words and phrases and passed them back and forth. They made up titles for any given moment and kept this hoard to themselves, like the shells and seedpods they picked up and stuck in jars on the back porch. They had a shelf of collections—screws, burnt matches, spent shell casings, twists of copper wire.

They were friends that way, during summer visits. Until a sequence of human messes, wipeouts, obliterations—a war and wounding, death, a woman alone and hating it, a knife, a lover's guns, crossfire—brought the girls to live with Aurora. They didn't want to be hers, or anyone's.

So they walked these days more than they read. Aurora walked them up and down corn rows until their forearms were cut from the leaves and they all felt crazy. *The Rise and Fall of Monoculture,* Laila said. Not long after, Ginny swiped a beat-up canoe from a dump in the woods. She said maybe they'd better get out of the fields and onto water.

* * *

The air simmered. Laila rolled her shirt halfway up her chest, then peeled it off, too, folded it and sat on it.

"Good. You girls are wiser without your shirts," Aurora said.

"*The Three Wise Women Return,*" Laila said. "Having gathered berries and nuts."

"A heap of nuts," Ginny said.

Although they weren't returning. They hadn't been anywhere,

and they weren't going anywhere.

Laila steered them into a cove, into deeper water where cattails couldn't root. They drifted. It wasn't a plan, but they eased toward land.

Spotted jewelweed lined the shore, its flowers lipping the water, and beyond that stood a thicket of prickly ash.

Ginny pointed. "Hit right there." She paddled hard to give them momentum, then lifted the paddle out of the water and ducked down.

They swept through the flowers. Aurora and the girls leaned themselves forward and pushed, to grind the canoe into mud, into the prickly ash.

"Get your shirts on," Aurora said. "Those thorns'll tear you up."

"No way," Ginny said. "We can duck under, and then it looks clear."

"Thorns tear your shirt anyway," Laila said.

Ginny was already out of the canoe. Laila threw their backpacks under the shrubs and jumped out. Aurora stood, and the canoe rocked.

"Wait a minute," Ginny said.

The girls beat down some of the jewelweed stalks, a mat on the mud.

"OK. Here you go," Laila said.

And like the old lady she was, the royalty she was not, Aurora stepped out of the canoe, one hand in each of her granddaughters' hands.

The girls charged ahead, slinging their sacks overhead and side to side, breaking down thorny branches. Aurora strolled without stooping, through that archway into the clear, into the underleaf shade of the woods where the air was green and watery. The cool was a shock, and she breathed it in, dark oxygen through her lungs, into her brain, a basement smell, through a door, a step down into a dirt-floored root cellar.

The girls slung their sacks on their backs, picked up dead branches and smacked their way along. Under their feet, sycamore leaves as big as shirts lay heaped on the ground, last year's fall. The girls kicked them around, flipped them over.

They danced like that for a while, snapping twigs, jumping into and out of sun blotches. Aurora trailed along, camouflaged in the mottled light. If the girls turned around, they might not even see her, stopped in front of a sycamore trunk, part of its shifty grays and greens.

Midway through the woods, in the middle of the tallest sycamores, Aurora felt a dry gust of air, like the wind shift in a new weather system, something driving it. Engine noise. Machinery. She stopped, turned left and right, radar head, to listen.

But something else, overhead, more strange, interrupted— drawn-out choking sounds.

Ginny yelled, pointed up. She was scanning the sky with binoculars. Several large shadows passed—great blue herons, their bodies angled between the high branches, their long legs lowered for landing.

Aurora could see it—they'd crossed into Birdland. At the tops of the sycamores, dozens of nests, stick-heaps, were wedged between branches—a heron rookery—stashes of gatherers' bundles.

"Wild!" the girls said. They mimicked the herons' guttural calls, and like girls in the wild before books, before words, they swung out their arms and shadowed the shadows.

⋆ ⋆ ⋆

Aurora leaned against the trunk of a tree. So they had got somewhere.

The herons circled lower and glided down, intricate limb-work and puppeteering, for stand-up landings on the sticks.

On the nests, they were immediately silent, great blue vanes, wings wide and sunning. Aurora shaded her eyes to watch, but the herons were stick-figures now. She couldn't tell bird from branch. She didn't care. She watched the tops of the trees for a long while.

⋆ ⋆ ⋆

One heron suddenly lifted off, one wing fold. Aurora turned her head and caught the engine sound again.

She'd forgot the girls, completely forgot them. She cupped her hands and called. No answer.

The machine noise rolled through the trees, engines revving up. On the other side of the woods was one of the dairy's cornfields. Somebody was driving something out there.

She called again.

Aurora couldn't move fast, but stepping away from the tree, walking under the nests, she was light on her feet, on her long bird-bone legs. She felt a feathery flesh on her thighs, on her forearms, her skinny neck. She was tall, and she thought the word *airborne*. She flapped her arms and the air pushed up against her skin, cool at the elbow bend.

Up ahead, beyond the rookery, rose a flat-top hill. From there she could probably spot the girls.

It was a climb. The side of the hill was steep, and the leaves kept slipping away beneath her feet, heavy in her shoes again. Her arms lost their lift, and she struggled to pull her legs up, step after step. She held onto shrubs when she could, grabbed the low branches, and pulled herself along. The engine noise grew louder, and it wasn't just one thing running, there were many machines, grinding, laboring.

At the top of the hill, up on the flat, Aurora stopped. She was breathing hard. She held onto the trunk of a skinny sassafras tree like a cane, like a parasol, and looked around.

Machine noise rumbled, she felt the vibration in her chest. The noise darkened the woods. Aurora squinted, it was hard to see. One sense interfered with another.

But in the distance, at the far side of the flat ground, Aurora caught a glimpse of Ginny sliding off the edge, out of sight.

Beyond that drop, beyond the darkness of the last trees, a rectangle of sunlight glowed and roared. The cornfield was gone, she could see that. The field was bare dirt, explosive light. Through the low branches, Aurora caught the motion of machines, swaths of yellow and dark cuts of brown—graders, bulldozers—a wall of dirt. A wall of dirt?

Aurora pushed with her hands in front of her, clearing the way, and walked in the direction Ginny had gone. The woods and even the leaf litter under her feet were blown away by the racket, she didn't notice the dark ferns that brushed against her ankles, she didn't hear birds, she couldn't. She covered her ears, but that was no help.

When she reached the edge of the woods, her legs slid out. She must have shut her eyes because when she opened them, she was sitting on the ground, her knees bent over the edge of a ravine, her feet dug into the slope.

She could see, she wasn't blind, but this was no place she knew. She might as well have been dropped at the edge of a desert, a dumping ground in the Sahara where corrugated tin and hides were shelter, or mirages, where there was no water, where animals crumpled and died, where their carcasses were home to small-boned boys, or worse, when it got worse, the valleys were strewn with blood and bone shards, exploded armor, tar, axles.

The noise rolled over the wall of dirt and across the ravine. Aurora sat where she was—and there they were, Ginny and Laila, far down the slope, on their knees. Their backs shone with sweat.

The girls were holding their noses with one hand and holding long sticks with the other, poking at bodies, mounds of dead bodies of animals slightly smaller than themselves. Foxes. Coyotes, maybe. It *was* a desert. The girls pushed at the animal bodies, which looked more liquid than solid, in motion with swirls of maggots, Aurora could see the flow, and patches of fur sliding. Longer legs stuck up like posts in the heaps. Dead calves? Hoofed, cloven feet.

The girls knelt in the midst of a tangle of hoops and twists of barbed wire, they could be slashed, they could slip on the dead animals and even bleach wouldn't help, there were rolls of rusted chicken wire, old plastic sheeting, paint cans, a pyramid of beer bottles, junked trucks somebody pushed over the edge, rusted fenders, tires, heaps of dumped rocks overgrown with wild grape vines.

How could they breathe?

She knew they were breathing.

Another blast of machine noise crossed the ravine. Aurora felt the wind of it and looked up. An earthmover, very close, came around the base of the wall of dirt. On top of the wall, a bulldozer pushed a bladeful of dirt over the edge and churned up dust from its tank treads. At the rim of the wall, the bulldozer turned away and drove down inside, reappearing on another wall of dirt. Down one side, up another. Aurora watched, up and down. She counted four walls of dirt. A pit. A manure lagoon.

They'd stuck it back here. Another dump. Behind the woods, behind the marsh. Nobody could see it. But from the air! And she saw it, with heron eyes from Birdland or wherever she was in whatever wing-it territory, she eyeballed the length and width, the acres, the millions of gallons. She could see past the walls of the pit, she knew the slope and the depth.

She shut her eyes, measuring, multiplying, multiplying.

When Aurora opened her eyes, Ginny and Laila were picking her up, her arms were draped across their shoulders, they were lifting and turning her away from the wall of dirt and the dump.

"It's OK. We've got you," Ginny said.

"*Carried Away*," Laila said.

Aurora's feet were still turned downslope, but the girls didn't notice.

Her feet dragged that way, backward behind her.

The girls smelled rancid.

Aurora thought if they dropped her, a bag of bones, she'd get up, fleshed and feathered even. She'd take off. She could go AWOL, leave the girls, why not, they're breathing fine, but then again, she knew how it happened in nightmare, you never got to the water, you flew over a fence, crossed into another country, and there you were in the same place with the same guy stabbing the same knife in your neck.

DOWNSTREAM

"I've had it with cow towns," Tatia said. "This place is the past tense."

So it was.

We were out at the stream on my farm last summer, in fine ripply shade. Tatia straddled an oak tree fallen over the water, and she looked off through the woods, right through the pawpaw leaves fanned out near her face.

"I'm going thataway," she said. She was pointing east.

The water of the stream flowed clear, right down to the mottled rocks, blue and brown, and the fine gravel fingers of the streambed. The stream was spring fed, with no cows, no cowshit, upstream. I was lucky to have good water. In the swirls and eddies, we'd often found freshwater clams, and under the rocks, mayfly larvae and stoneflies with forked tails.

"If you move downstream, it won't be like this. It gets ugly," I said. "Foul. Things don't get better."

Tatia didn't care about water quality. She was on her way out. She'd moved here a couple of years ago, into the rented house across the road. She read books and she edited manuals for technology companies. She had it backwards, I told her. She could move downstream six times and never get closer to anything.

But that didn't make a dent. With her shiny black hair, those fine long toes, you'd think she'd have wanted some softer ground, an upstream pasture, or forest.

But Tatia thought the sea was kin to her, and she'd somehow been stranded up here where water was just getting started. She never said ocean. She said *sea*. *Kin*. The *sea* was *kin* to her.

Some people don't talk enough to their neighbors. Their language suffers, they are so literate. Tatia couldn't think what to call ordinary sounds, like the smack of a hand on a thigh. She wouldn't make up words, she thought it wasn't right. That's the smarts that got her as far as Toledo, the last I heard. She has a thousand more miles to the sea, downstream—though it looks on the map like up—to Niagara Falls, and Montreal, and the mouth of the St. Lawrence opening wide into the Atlantic.

★ ★ ★

Marvin says that's how it goes, reverse migration, now that we've got to the end of the line. He's stopped along the road where I'm sampling water from the ditch. He wants to talk about Tatia.

"I'm sorry to see her go," he says. "She was sweet." He means unlike me, and he says it. "Unlike you," he says, and pats my arm.

"The only clean place to go is outer space, but who can afford it? Not even me," he says.

For a dairyman, Marv has the perfect name, Rindhoof, and he's worked it into the logo on his cap—*Rind* ribboned around planet Earth, *hoof* trailing into space. Tatia liked to drive to Ohio with him, to the bar in Archbold, the nights he could get away from the house.

Oh, cow town it is. Marv in his white truck is duke of the fiefdom. He isn't as tall as his brothers, the accountants. He looks a lot like the ag engineer Rickie Tea from Michigan State, with a small round face and blond hair in a crease across his forehead, like an eyebrow. When they go around checking manure draglines or drainage pipes, they look like twins, walking in step, with the forehead half frown, half goofball smile.

Tatia hated the boomtown spurt the big dairies brought, with the miles of steel confinement buildings along the dirt roads. She liked all the men, though, and got them to talk, no matter how pissed they were when they sat down at the bar. They told her about their wives, about problems with plumbing and liquid manure, about trouble with those goddamned water testers, and Tatia would come back, sit on the fallen-down tree, and tell me.

She had no principles. She liked men with problems, and of course there are plenty of them, upstream and down. I'm sure she's found more, wherever she is.

Tatia and I disagreed about a few things, but not much. I thought certain men should not be allowed to touch taxpayer money, for instance.

"Just men?" Tatia said, "Which ones?"

"Well," I said, "Dairymen. Maybe the road commissioner, with that dull skin—"

"I *know* him!" she said. "He told the road crew to chop the flowers in Bobbie's front yard. You know her? The name Bobbie got to him. It's one of those things, he said, who knows if it's a man or a woman. If it's Bobbie, chop the goddamned flowers, pal. Chop the goddamned shrubbery. Say it's for safety. You can't be a road commissioner and not worry about safety."

"You've had some conversations," I said.

"You said it," she said.

But we did disagree about travel, and now that she's gone, I don't think she'll write. It's been months since somebody saw her in Toledo, and I haven't heard anything. She had the idea that I was too settled. Stuck, in fact. In mire, she said. She laughed at the black rubber pig boots I wore, and she'd kick out her legs once in a while, to show off the slipperette sorts of shoes she always wore, even walking in the woods, in mud. Tatia said she was a dancer for a few years, and I believe it. She had the posture, and could cross the stream without getting her shoes wet, with one long leap, her legs in a split.

Marvin talks to me because he wants to talk about Tatia. He hasn't heard from her either. We sit on the running board of his truck, while I take off my rubber gloves and fill out the sampling log.

He says, "She had no animosity." His voice is formal, with a slight accent that slows him down, a pointer on each syllable. He's from the Netherlands, drawn across the sea to the big dairy cash flow.

"She didn't care about the water," I say.

"Because she did not drive around day after day, causing trouble?"

"She didn't care about *this* water," I say. "She wanted the ocean."

"Well, people like her should get what they want," he says. "I'm glad she got the chance. I would go to the ocean, too, if I could," he says. He stands up and steps out into the sunlight.

"So go. You've traveled." I take the sample bottle to the car and put it in a cooler in the trunk. "Do you miss the sea?"

Marv turns my way, and he looks over into the ditch, at the brown water. "You can wait for goddamn-it forever," he says, slowly, "and this water won't turn blue. If you want to see blue water, go to the ocean."

"I'm not going anywhere, you know that. I asked about you."

"I have thought about it," Marv says. "If the government buys me out, hey, I would get out for sure. I get not a thing but grief from you people."

I take out the dissolved oxygen meter with the twelve-foot cable. "You want to test your water?"

Marv takes the meter, and I drop the cable into the stream.

"Watch the numbers drop," I say, because I know they will. "If it gets below 5, it's no good for fish."

He stands there very quiet holding the meter, and he holds it gently because it's digital and he knows about delicate technologies.

"What is it?" I ask.

"3.3," he says.

He respects the machinery, and he shakes his head. "I do everything right," he says. "I do what I'm supposed to do. Some of this stuff drains through the soils. This could come from septic, you know."

"Then that's yours, too. There's nothing between us and your place."

"Tatia had the idea she was a saint," Marv says, out of the blue.

I just sit down where I am, on the slope of the ditch. Marv hands me the meter and sits down, too. The ditch is a pretty place when you sit in it, with lobelia blooming, and clumps of sedges near the water.

"She told me one time she was pure in heart and had never done any wrong," Marv says.

"I never heard that," I say. "I heard plenty of other things."

"I know, I know, she slept around, a little. But she kept above the fray, you know what I mean?"

"No, I don't have a clue what you mean, Marv."

"Well, she wasn't attached to things, she didn't give anybody a hard time or make them feel guilty now, did she?"

"She laughed at my boots," I say.

"But it didn't make you feel bad, I know, that's what I mean. Not like the tack you take."

"I do not take a tack. I sample the water, I tell people about it. Don't you want clean water?"

"I'm talking about Tatia here. She tread lightly, you know what I mean?"

Now there I do know what he means. Tatia would have said those words, too, *tread lightly*. She who treads lightly is kin to the sea.

"She took me to Harrison Lake one time," Marv says. "At night. We left the bar, and she drove the truck. She told me I ought to get outside more."

Marv leans back. He's halfway down the slope, and his head rests against the weeds. "It's true, I don't get outside much. I manage more than farm."

"You could downsize," I say. "Go back to real farming."

He ignores me. "The park was closed, but we drove in anyway. She pulled off the road, and we walked up this grassy hill, like a dam maybe. She pushed me down on the grass, flat on my back, and came at me with her mouth open and gave me this big mouth kiss."

Marv shuts his eyes as he tells this, going back into the dark. "A big kiss," he says, "like the sky came down and just laid itself on my face."

"The kiss of a saint?" I say.

"Yes, it was," Marv insists. "I am only telling you because she is gone, and because she was better to me than anybody. She didn't ask for anything. She didn't want anything."

"Then why did she leave?"

"That's just the point. She wasn't attached to anybody. She could go where she pleased. She was here, now she's someplace else. It doesn't matter to her."

We watch the water for a while. Some dead snails float by. He's right, Tatia is probably kissing some guy in Cleveland, or walking along Lake Erie, her feet hardly making prints in the sand.

Marv shakes his head. "These days, it is like waking up every day," he says, "in a less favorable climate."

Although he's soaking sun today in this climate, his face catching the full brunt of the heat, and a bird is nearby, singing like crazy.

* * *

Tatia had a routine in the mornings. She woke at sunrise and took a cup of coffee outside, even in winter. She sat in a deck chair facing east. I saw her sit there in the rain, in the wind, it didn't matter. On her refrigerator, she'd taped a sheet of paper listing the hour and minute of all the year's sunrises.

"It's what I do," she said, "to get started. Work is easy after that."

Through the day, she didn't open the door. Maybe she worked steadily at the computer, maybe she watched the soaps, I have no idea. But evenings, she'd take off. And weekends. That's mostly when I saw her, on Sundays. She'd walk across the road after lunch, and we'd head back the lane to the woods. She said her next job would be some kind of manual labor. She was sick of working inside. She talked about Marvin, or one of the other guys, not in a gossipy way, just the facts. "Marv says the pit's full. They'll be spraying manure downstream of you this week. Anders wore that white cowboy hat again, he doesn't get it. He thinks it's the American thing to do. He had a spill from the Terra-Gator hauling down on Packard. Did you know about that?"

Tatia heard everything that was going on. By Monday, I had a pretty good idea where to test the water.

* * *

"She had such a clear mind," Marv says. "No clutter."

"If nothing matters to you," I say, "I guess there's no clutter. Is that a clear mind, or empty-headedness?"

"Now that is not fair," Marvin says.

I don't like the sound of it either. I just hate to agree with him. "No, I know, she's smart," I say, as apology to Tatia. "She always says what's on her mind."

Nearby, the bird is still singing, a song sparrow, what a racket. There's a breeze whipping the tall grasses above us. Purple loosestrife is taking over where it shouldn't be. The water smells. A sheen of something floats on the surface, with brilliant paisleyed colors, iridescent. Yellow bur-marigold is booming, big bouquets, at the water's edge.

I take the DO meter back to the car. Marv doesn't budge. Sprawled on the slope of the ditch with his eyes shut, he looks like somebody knocked him there, knocked him out.

"Rindhoof!" I call down. "I'm headed to the stream on the other side. Want to see what's what?"

He doesn't move.

"Marv, are you okay?" I slide back down the slope to check.

He lifts his head because I guess he has to sometime, and he sits up. His feet are almost in the water.

"We're the grunts," he says. "That's all there is to it."

And now he's done it—stuck me and him together forever in the same ditch. While Tatia in her sainthood rises far beyond us, leaps skyward someplace, into the blue.

"There's nothing we can do," he says, "to bring her back."

2 | FIELD NOTES

THE FANTASY OF THE CLIP ART FARM

Farm is a four-letter word. Right up there with *flag, love,* and *home.* Any kid can draw *farm* with a few crayons—a red barn, a silo beside it. As easy as *love,* a heart, and *home,* a house. But these days, in many rural communities, *farm* is an f-word, too. With livestock factories reshaping agricultural areas, *farm* has fallen hard—plummeted—from the sacred to the obscene.

The livestock operations that surround my Midwest town, Hudson, Michigan, still call themselves farms. Most are dairies, and they're all huge, all built within the last few years. In the language of law, they're CAFOs (Confined Animal Feeding Operations) with more than one thousand "animal units"—that is, seven hundred or more confined cows—and open-air waste pits that hold millions of gallons of liquefied feces and urine.

Like the hog CAFOs in North Carolina, Missouri, Washington, and chicken operations all over the place, they're animal factories that from the air resemble airplane hangars, the largest constructions on the new rural landscape.

From the ground, the buildings are long, low steel constructions, some of them a quarter-mile long. When I showed a picture of one of these places to my Mennonite cousin in Pennsylvania, she said it reminded her of a Mennonite church, plain and simple. She got the up-front message—this is a clean place, orderly, unadorned. CAFO dairy design is spare and severe—white siding on the long walls of "freestall" barns where the cows are confined. If the milkhouse faces the road, maybe they'll add a few decorative bricks or tiles as façade. At the largest facility (four thousand cows), just opened for business south of town, an entry drive leads past a circular pool that looks like it might have a clean-water fountain upspurting, come spring.

Out back, however, out of sight except from a plane, are the pits, lagoons, and cesspools of millions of gallons of untreated animal waste.

What you *never* see at CAFOs—not from the road, and not even from the air—are the things most people picture when they think of "farm." At CAFOs, you will *not* see

cows outside grazing.
red, hipped-roof barns.
pastureland.
upright silos for storing feed.
straw as bedding.
windmills.

Yet these elements persist in ads, brochures, the imagination—

the full iconography of the myth of the American farm. The red-barn-&-silo-background with green-pastureland-foreground sells cheese and milk on TV to this day. These features appear without fail in the Clip Art images Microsoft provides as "farm," including this image—

—chosen by the Michigan Department of Agriculture a few years ago for the cover of its Web publication *Generally Accepted Agricultural Management Practices*.

When designers draw an Ag image, they cluster and stylize the same farm elements that kids draw. And, more important, when the lobbyists for industrial agriculture, like the American Farm Bureau Federation, argue for no regulation of CAFOs, they talk about family farms, the farmstead, and conjure up the same small-scale Clip Art image.

The Clip Art Farm—red barn, hay mow, silo, windmill.

In fact, the Clip Art Farm elements are 30 to 130 years old, anachronisms in most rural areas of the Midwest. The red, hipped-roof barns, for instance, like the one with haymow brimming in the Department of Ag image, were constructed around here in the late 1800s, with some newer ones built through the 1930s. Seventy and more years ago, loose hay for feed and straw for bedding were lifted into the mows by a system of knives and pulleys. In winter, livestock were housed in the lower levels of the barns, and hay could be forked down to them through trapdoors or chutes. In summer, animals fed outside on pasture.

Windmills date from the land-clearing settlement days of the nineteenth century, when they served as the energy source for pumping water. Almost all windmills ceased operation when electricity and electric pumps came to rural areas here in the 1930s.

The windmill is a fine design element though. It's hard to give up! And easy to oversize as well, like a sun shining over the farmstead.

Silos—those erect ones in the Clip Art Farm—replaced the haymows for storing animal feed. With silage choppers and power augers, feed could be mixed, lifted into silos, and moved back to stalls without much handwork. Silos were constructed relatively recently, from the 1950s to 1970s especially, and smaller farms still use them.

In the new rural landscape, however, silos do not appear; they can't store the quantities of feed required in confinement facilities. CAFOs use concrete "bunker" silos—they just call them bunkers. Bunkers are excavations with concrete walls on three sides. Hay, corn, and other feed is chopped, then dumped, and hauled in and out with front-end loaders. Bunkers are usually covered with black plastic held down with tires, one of the low-tech details at these "state-of-the-art" facilities.

The *lowest*-tech feature of CAFOs—the one that least lends itself to image design, that most contradicts the Clip Art image— is the effluent holding pit. Manure "lagoons" are open-air, earthen or concrete-lined pits that hold millions of gallons of animal waste, contaminated stormwater, and milkhouse wastewater.

Cows used to graze in fields and deposit their waste here and there. Not here.

Here, all of the cows are confined year-round, and all the waste is collected in one place. The CAFOs use scrapers and huge quantities of water to pump all feces and urine into pits, lagoons as large as some lakes. Eventually the liquefied manure will be trucked or pumped or injected or "spray-irrigated," as they say, onto fields.

Liquified manure stinks. It is not the sweet smell of old-time manure, mixed with straw, composting into soil. Liquid manure steeps for months in the pits; it is potent with pathogens, hormones, antibiotics. Sprayed onto fields, it can flow on top of the ground, or seep through the soil with its contaminants, entering field drainage tiles and streams.

Across the country, agricultural runoff of soils, pesticides, and manure is the major polluter of rivers and streams. The EPA estimates 40 percent of the waters of the United States are too polluted for fishing or swimming. More than half of that pollution—60 percent—comes from agriculture, and with the liquefication of manure, CAFOs have become the principal contributor to agricultural pollution. More than 2 trillion pounds of animal waste per year are applied or sprayed onto soils.

Even though alternative waste systems exist—composting, pelletizing for dry application, sterilization and drying—industrial

Ag facilities here have resisted any requirement for dry systems or any waste treatment (systems even small campgrounds are required to install).

It is understandable why state agencies and their research arms, the land-grant universities like Michigan State University, have not jumped in to study watershed risks or public health issues. Big research funds come from Big Ag, after all, not community environmental groups. "Sustainable" agriculture is a subdivision of most university Departments of Agriculture, as if only some, not all, agriculture can be sustainable. In fact, here in Michigan, the state Department of Agriculture and MSU's Ag programs have been instrumental in promoting industrial livestock facilities, in particular the freestall barn and liquefied waste systems for dairies. Unfortunately, MSU's history of denial of environmental risk is notorious.

Decades ago in *Silent Spring* (1962), Rachel Carson recalled MSU's spraying of DDT on the campus, obliterating the robin population, "in spite of the assurances of the insecticide people that their sprays were 'harmless to birds.'" She also recounts the Michigan Department of Agriculture and USDA's use of pellets of aldrin, "one of the most dangerous" of insecticides in a dusting campaign over Detroit. Officials declared that "the dust is harmless to humans and will not hurt plants or pets," although the bird population was "virtually wiped out," and cats, rabbits, and fish sickened and died in great numbers.

That's all in the past, of course, but so are windmills. The legacies of wish and denial complicate current policies and research.

Agriculture has a profoundly compromised past in this country, where denial and the arguments of economic necessity have supported a range of injustices and degradations, both human and environmental—from the use of slaves in southern agriculture; to the draining of wetlands throughout the Midwest; to the damming of almost every river for irrigation of farmland; to the ongoing irrigation of semiarid lands, depleting aquifers, for farming. Through it all, agriculture claims it *must* do what it does; farmers know best what is best.

The moral and essential claim of agriculture, to do whatever it wants because it *must* or face ruin, has rarely been questioned by Americans. The myth of the *farm*, envisioned in the Clip Art Farm, holds powerful sway, as if the goodness of a creative God and all His righteousness had been shifted to human farmer hands. Who are we to question Agriculture, those who plant and grow our daily bread, who feed creatures to be our meat? Adam may have been booted out of the perfect farm, the Garden of Eden, but the Clip Art Farm constructs itself as alternative (a fallen farm, maybe, but in a farmer's control—finally).

Farm-friendly Congress and most state legislatures listen to Big Ag interests with devotion and little skepticism. The Farm Bureau is one of the most powerful lobbies in Washington, consistently appearing in *Fortune*'s "Power 25" lists. In farm bill discussions, concessions to Big Ag on "conservation" measures have lead to *more* funding for CAFO lagoon construction, rather than funding for smaller operations or for alternative dry-waste systems.

It's certainly no secret that farming has changed, consolidated, and concentrated. Plenty of attention has gone to the corporate guns—Cargill, ADM, Monsanto, Tyson, Smithfield—in articles like William Greider's "The Last Farm Crisis" and widely read books such as Richard Manning's *Against the Grain,* Eric Schlosser's *Fast Food Nation,* and Michael Pollan's *The Omnivore's Dilemma.* It's no secret the food system has been industrialized, raising global concerns and frequent debates over farm subsidies, raising national concerns about antitrust issues, the "lawless" market economy, the promotion of fast-food calories, and the obesity epidemic.

At the local level, rural landscapes like this one have consolidated, too, both visually and economically. In the last decade, this area has been dramatically recast and reshaped, from many small and varied farm operations to several huge dairy CAFOs, which own or rent more and more of the landscape.

"Rural development" is the catchphrase for this agricultural sprawl, with tax incentives and grants, giveaways for road reconstruction when CAFOs' constant trucking tears up pavement, and massive farm subsidies to industrial Ag facilities.

You hear it all the time around here, an outrage to some but a boast from Big Ag: farmers know how to farm the government.

It's no surprise that whatever Big Ag interests touch these days, it carries a taint in communities like this, a kind of unnatural stench. Even Farmland Preservation, a program to purchase development rights (PDR) and preserve farmland—a program that environmentalists as well as farmers usually support—has been sullied here by substantial PDR payments to the largest Ag land speculator. It's a system that has promoted Big Ag ownership of thousands of acres, across several townships and two county lines—a feudal shift, with the domain control of agricultural lands in a few hands. PDR money can cycle back, buying up more small farms—more land for more PDR money, or to sell or lease to the CAFOs.

"Preserved" agricultural land can be used in *any* agriculture-related way—huge buildings, lagoons, Ag processing plants.

So instead of farmland being preserved as open fieldspace, grain and greenery, which people imagine surrounding the Clip Art Farm, it in fact can be fully *developed* and industrialized, with factories and effluent waste pits.

The private interest in the profitability of industrial agriculture often overwhelms the public interest in environmental protection. When people around here—small farmers, teachers, business people, lifelong residents—express concern about the CAFOs' impact on water quality and the risks to watersheds, they are often dismissed by Ag interests as outsiders. Of the local CAFO issue, one state commissioner of agriculture recently said it was simply "a conflict between farming and those who desire to move from urban to rural areas."

I've lived on farms all my life, as a child in Pennsylvania, and in this area of Michigan my whole adult life; I've farmed hay here, restored wetlands on my farm. Still, from the Big Ag angle, *everyone* living in rural areas, no matter how long they have lived there or even farmed there, is seen as encroaching on (their) agricultural zone.

The property rights of neighbors and the rights to a healthful environment are subsidiary to industrial f——ers' rights. Anything zoned as Agriculture is treated by industrial Ag as a safe zone, for the practices they please, no matter who is affected, displaced, harmed, or sickened. Environmentalists are accused of caring more about water than people and dismissed for that; but even the interests of small farmers are ignored, or niche farmers like nursery growers, truck farmers, orchard growers, organic farmers. If you're not Big Ag, you're nobody. And these privileged rights of industrial agriculture are certified in many states by law, with Right to Farm legislation, and in some places with so-called Agricultural Security Zones, where farmers cannot be cited for "nuisance" activity (emissions, noise, dust).

Around a small farm, of course, freedom to operate the farm business makes sense and is rarely a nuisance by any definition.

Around a CAFO, however, nuisance is a mild word for the intense stench, ammonia and hydrogen sulfide, fecal dust and particulates, truck traffic, all night floodlighting, manure sprayed on soil, and manure sprayed on snow. Rural neighbors often have to live with worse than any city conditions of air and water pollution.

Most neighbors finally give up, give in, and sell—if they can. Rural communities suffer division and real loss, as people near CAFOs move out rather than hazard the respiratory complications that are common among CAFO workers and families. My family doctor lived around the corner from a site bought by the largest CAFO developer. Concerned about the emissions and his family's health, he moved away when construction began. His family had lived and farmed in the area for seven generations.

The risk to public health and the health of watersheds is no small concern. Agricultural pollutants and pathogens, excess nutrients, and antibiotic-resistant bacteria threaten water quality across the country. In 1993 in Milwaukee, 111 people died in a *Cryptosporidium* outbreak, and 403,000 people were sickened. Maryland's Eastern Shore has had serious problems with the toxic microbe *Pfiesteria piscicida* causing major fish kills. In Walkerton, Ontario, in 2000, 7 people died from *E. coli* 0157:H7 in public drinking water supplies. These contaminations have all been linked to pollution from livestock operations.

Even so, at township and county meetings in my area, you'll still hear denials of any farm fault. My township supervisor, who raises calves for a CAFO, said Walkerton just had a problem with wells.

In this fantasy, every farm is a Clip Art Farm. It's a good, a sacred place. A Clip Art Farm cannot pollute.

Farms, even the factory f——s, are hallowed and haloed.

So in Michigan, CAFOs continue to dig waste pits, and then more and deeper pits, even where the water table is near the surface. Clay lagoons leak, concrete lagoons crack.

It's the oldest lure in America—bigger is better, size matters. In the USA, CAFOs—the biggest operations around—take the bait. It catches them hell, of course. Neighbors get ugly when they can't breathe in their backyards on sunny days, when their kids can't picnic or play outside. When they can't swim in the creek. When their property values evaporate in the stench.

Around Hudson, environmental groups have popped up, a watershed group; many business people, small farmers, and more recently church groups have been working for regulation and inspection of these operations. Until recently, Michigan had no permit system, and now it's a fragile, shifting set of rules. There's no routine inspection of lagoons, no water monitoring of pipes that drain manure fields, no monitoring of manure-sprayed air. Nobody here has measured the hydrogen sulfide emissions from manure application near a day-care center. CAFOs still spray manure on frozen ground, on tile-drained fields. If CAFOs are following their manure management plans, they're "in compliance," says the Department of Agriculture, no matter the mess, the stench, the suffering of neighbors. That's f——ing these days! Generally Accepted Agricultural Management Practices (GAAMPs) are voluntary "guidelines." They're "recommended."

CAFOs don't even need building permits—they are "agricultural" facilities and are therefore exempt from such regulation. Milking systems are inspected, but there's no regulation or inspection of the tail-end systems. One CAFO operator needed a permit from the township for a backyard swimming pool; he didn't need a permit from anybody to construct his 1.25 million gallon waste pit.

A surprise EPA inspection of one facility here found floor drains directly, and illegally, connected to field tiles, which discharged into a nearby stream. Most of the Hudson-area CAFOs were built by the same Ag construction company—a company spun off with MSU Ag expertise and affiliated with the largest CAFO development corporation. So there's a good chance other facilities have direct tile connections to drains and streams.

But who knows? Nobody sees the plans or approves the design; nobody inspects the construction.

The national belief in the unassailable virtue of all farmers, like a dreamy belief in the virginity of all maidens, persists in some places, although probably not for long.

Ag people and politicians have to look way back, back to Neverland, or they have to be as nostalgic as Clip Art designers, to call farmers "the true conservationists," as both Gore and Bush did in their presidential campaign of 2000. There's a strange fear, like the fear of loss of faith, to admit that farmers are fallible.

American agribusiness, Ag producer associations, and Farm Bureau, in particular, lobby hard with the fantasy of the Clip Art Farm. Without it, they'd have to admit that farming's a business, and a CAFO is an industrial facility. They'd have to acknowledge farmers are human beings. And need, like the rest of humanity, not just voluntary guidelines but laws.

Infallibility is a difficult claim to make for anybody. Farmers have managed to walk on water for about a hundred years—literally, even, in Michigan, by draining wetlands and burying streams in pipes. But that's not much more than a trick, and they're being outed. They're just business folks working the resources—some decent, some money-grubbing, most of them no more committed to preserving the environment than mining companies topping the mountains in West Virginia.

The nationwide proliferation of animal factories has turned many livestock farmers into operations officers, business managers. They may wear the John Deere cap, but they mostly don't drive the tractor. These f——ers don't really farm.

Across the new rural landscape, that's the resounding crack you hear in the old mythology.

When operations get CAFO-sized, farmers can't work all the fields, can't know all the ground directly. They buy up other farmers' land, they bulk up their holdings. They hire low-wage laborers or subcontractors like the Tail-End haulers with their tankers to truck manure to remote fields. Chances are those drivers have never seen the fields before and certainly never tilled them. One CAFO hired a trucking company to haul loads of thick waste from the bottom of a lagoon, load after load dumped in a field. That's fertilizer? That's the "agronomic rate of application" the guidelines recommend?

Too often, it's plain old illegal waste dumping.

One CAFO operator, relocated from the Netherlands, wrote with enthusiasm about how farming was finally a living rather than a way of life. And why not? Lots of folks like to have a life as well as a job, go on cruises.

But as managers of industrial facilities, they hang on for dear life to that word *farmer* and the fantasy of the *farm*.

The myth isn't flimsy. It has two powerful drivers: first, there's the "farm" (unlike factory) relief from regulation, and second, maybe even more powerful, "farmers" receive farm subsidies, a welfare system that dwarfs any assistance offered to anyone else in this country.

"Farmers" receive state and federal subsidies, loan guarantees, and outright payments that nobody else could dream of receiving. Payments that few taxpayers, if they knew, would dream of supporting. The bigger farms get, the more industrial their facilities, the more support they get. The numbers add up.

Farmers don't like welfare, of course. Welfare mothers?—Slackards! Abusers of the system.

Welfare farmers? Oh, no, no such thing! In spite of billions of dollars for farm subsidies: $165 billion from 1995 to 2005.

Food may be cheap on the shelf. But look where we pay for it: milk price-support money, economic development money, Environmental Quality Incentives Program (EQIP) grants, "conservation practice" money (lagoon, manure injector subsidies), production flexibility contracts, market loss assistance, loan deficiency payments. The list goes on.

In Hudson, the biggest Big Ag guy received a total of $923,980 in USDA farm subsidies for the years spanning 1996 to 2000, with huge increases each year. In 2000 alone, he received federal checks for $300,473. And that number doesn't include the PDR check he got for $794,957 from the state Department of Agriculture through Farmland Preservation. More than one million dollars to one f——er in one year. Elsewhere in the United States, farm subsidy payments reached the multimillions for individual producers. Cheap food?

Subsidies and "payments" have become so dear to the Ag industry that it can't see those checks in the mail as welfare checks.

No, Big Ag says, a farmer's farmwork on a farm is good, even godly. Patriotic. Farming is in the economic interest, in the *national interest,* in the global interest.

And if anybody suggests otherwise, Ag people call in the grand defense. It's the same back-at-you from MSU Extension, Farm Bureau, CAFO operators, legislators—"If America wants cheap food, . . ." "If you had to pay what food costs, . . ." or "If we're going to beat world hunger, . . ." and, as big as it gets, the first and last word:

"If we're going to *feed the world,* . . ."

Well, why should we feed the world any more than we clothe it or shingle all the world's roofs? Food is a business, one business, in the world's survival.

You'd think farmers were all Jesus, with the loaves and fishes! Not human beings with self-aggrandizing wishes.

Other people in the world are willing to pay more for food to protect their water. Here CAFOs continue to overbuild and overproduce at great risk to the environment and at great cost to taxpayers—the hidden farmer-welfare cost of food.

Because of water quality concerns, the Netherlands bought out many dairy operations and limited the size of others. Hundreds of Dutch farmers have relocated in the United States (in Texas, Idaho, Washington, Indiana, Ohio, Michigan—seven of the CAFOs near Hudson are Dutch operations) and operate facilities ten to fifty times larger than those they left.

"It's been a steep learning curve," one Ag official admitted at a local information session.

That's f——ing these days! On-the-job training in Hudson, Michigan—with fifteen thousand cows, fifty lagoons, and hundreds of millions of gallons of untreated, liquefied animal waste each year dumped on the soils and into the watershed.

Despite the CAFOs' wishful claim of "zero discharge" to surface waters, we've had more than 140 discharge "events" and violations between 2000 and 2006—animal waste over-applied, drained, dumped, or sprayed onto frozen ground, polluted liquid that flowed into drains and streams.

During one discharge, *E. coli* bacteria in a county drain reached 130,000/100 ml, a contamination level 130 times the acceptable level for partial body contact. The polluted water entered Lime

Creek, then Bean Creek, and crossed the border into Ohio, where the city of Archbold takes its drinking water, and on to the Tiffin River where the town of Defiance takes its drinking water, and into the Maumee River, outletting into Lake Erie at Toledo.

Water, air, soil—the natural resources of sustainable farming and all life—have been segmented, wrenched apart, in industrialized agriculture. Watersheds have been torn source from river from outlet, as if water didn't flow downstream and across all boundaries.

Around here, farmland is now a place to put a factory. Animals are units within that factory, confined day and night, birth to death—cow prisons, some call them. Cows are fed antibiotics, hormones, steroids for rapid and grotesque growth. Feed can be subcontracted, grown and trucked in. Waste can subcontracted, pumped and trucked out.

The Teamsters should move in and organize. F——ing's a trucking industry.

The road where I have lived for years is a country road, a dirt road like many in Michigan. Two years ago, the County Road Commission decided to widen and clear trees and shrubs from the entire right-of-way. They brought in huge hydrohoes to rip down the trees, including the two big maples in my front yard. When the road crews piled my trees and heaps of others on two-story-high piles and set them on fire, I called up the County Road Commissioner who lives down the road and who also happens to be a CAFO operator.

We stood in the middle of the road and looked at the machines, at the torn-up swath sixty-six feet wide, at the trees on fire, the smoke rolling, the mud, the whole mess. It looked like hell. And I said so.

"When I look down this road," he said, "I see a brighter future."

Many days now, when the CAFOs are emptying lagoons, and Tail-End tankers of liquid manure roar up and down this road, I count the trucks as they go by, every twenty minutes, hour after hour.

In many rural communities throughout the United States, this is the new rural landscape. If you're outside a city, look around—it's not hard to tell when you're near animal confinement facilities. *You will not see animals.* You will see bulldozers, tiling machines, the bunkers with black plastic, the Tail-End tankers. If you fly across the country, look down and notice the metallic glints, the clustered lines and rectangles, reflections off the long metal roofs of the thousands and thousands of animal factories our taxes support, the waste pits we subsidize.

And under those roofs, the factory meat and milk we pay for through the nose.

A brighter future? Contaminated water? Fouled air? We pay all this for what? Cheap cheese?

In the new rural landscape, it's a shock to see how far Big Ag has booted itself from the natural world. How far this f——ing industry has fallen.

LETTING GO
THE VIRTUE OF VACANT GROUND

Drive around southern Michigan these days, and you'll see the new look of farmland. It's wild. Unkempt. Downright gorgeous.

Where in the past you'd have seen a regimen of rows, now you see wild sprawl. Where there were singular crops, there's a riot of undergrowth, wind seeded. There is raggedness, mess, variety, mix. On its own, out of human control, farmland demonstrates an abundance—flourishing, bizarre, rank, twisted, vital. And the feel, in a way, is more urban than rural: boom and surprise and decay, all in one.

With the farm economy aggravated by grain surpluses, with corn and bean prices in steep decline, it's not unusual in southern Michigan to see a countryside visibly changed by the

Written in 1988, after the first Conservation Reserve Program and before the onslaught of CAFOs in the 1990s. This reads like a dream to me now.—JK abandonment of farms, or the abandonment of traditional field practices. A lot of places around here, everything's gone to seed. The federal Conservation Reserve Program (CRP) has opened up huge acreage to cover crops; and many farmers, on their own, have begun to let marginal fields lie fallow—just let them go. The gravelly hills, glacial dumps, the undrained low-lying clays— they're easy to give up now. It pays more, most of the time, not to till and plant them.

If you examine any square foot of unfarmed ground—a square foot that a few years ago would have contained some dirt and a couple of cornstalks—you will find this: quackgrass, of course, like a mat, and milkweed, ragweed, or pigweed, lamb's-quarters, maybe some bergamot. Weeds. All weeds. That is, wildflowers. Some native, some not. Nothing planted, nothing for profit. Nobody cultivates weeds. But in this economy—here is the virtue, to start—they cost the farmer nothing.

The virtues that follow, and accumulate, are more crucial, more enticing, in the long run; but the fact that wild plants are cost-efficient—no production costs, no harvest costs—is good enough to begin with. Farmers here, a few, have learned how to tolerate the mullein or goldenrod, rampant in a vacant field, and have figured out how to see something in the scene besides trouble.

To somebody who hasn't farmed, the scenery is unambiguous—it is spectacular. Weeds and native plants bloom in field after field of flower. This time of year, late July, the commonplace explosion of Queen Anne's lace—hand-size white bursts at the roadsides—proliferates outward into acres of old fields. It's a painterly, splotchy wash. Although bergamot and goldenrod haven't flowered yet, they've shot up to chest height, aromatic, feathery, so there is an expansion and bunching of foliage and vegetation, on to the horizon. In the most meager soils, chicory stalks, like stubble, appear—the flower bits of blue; and sweet clover takes over gravel slopes, a serious tangle of tough stem and small, indeterminate leaf.

Any parcel of marginal, uncultivated land breaks out in these stupendous varieties of weed, diverse but not as random as one might think. Without the inhibition of herbicides or the row arrangement of cultivation, plants emerge according to preferences for soil type, composition, elevation, moisture. By the looks of things—the colors, the textures—you can spot even slight gradations in slope, and all the shifts from sand to loam to gravel to muck.

Looking across these fields is an education in the designs of ease. In natural selection. Adaptation. Variety. In wildness. It's one of the plain pleasures of human vision in farmland: seeing something not human.

Gone wild, even the first year, a field becomes a thigh-high jungle. Anybody trying to walk through these waste places understands there is no vacancy, none, in "vacant ground."

By vacant ground, we mean, of course, unfarmed, uncultivated land. Uncontrolled by us.

Not farming marginal farmlands, letting some land go wild or lie fallow, was common farm practice, a regenerative practice, before the use of chemical fertilizers. Now, with some choice in the matter, in a redefined economy, farmers again can acknowledge the value of vacant ground, the virtue—why not call it that?—of letting go.

My concern here, the more I look around, is not agricultural but moral: a concern for how human beings use land.

Farming, like mining or house-building or any construction, is one way humans use land; it is no more "natural" a land-use than oil-drilling. So I am not talking about farming as pastoral, or profitable, or good for the soul. Not at all. Farming, here, is *provocateur*—it's made me think. And what I am thinking about is how human beings inhabit the earth.

★   ★   ★

I have lived in the hummocky glacial dumping ground of southern Michigan for most of my adult life, longer than I lived in the limestone-bedded and rolling farmlands of Lancaster County, Pennsylvania, where I grew up. And while Michigan isn't a wilderness, it certainly is wild, unruly, as scenery, compared to the farm landscapes I knew as a child. This part of Michigan—a good bit of it swampland when settlers first arrived, a place notorious for mosquitoes and bad air—remains, because of the problems with drainage and the more extreme ranges of seasons and rainfall, much less domesticated, much less rich than eastern Pennsylvania. Not many farms here have the neat arrangement of barns, with good paint, the encompassing fields and pastures, that show up on the postcards of Lancaster County. From a Michigan point of view, though—and that's what I have now—the pastoral aura, the precision, of postcard farms looks more like tyranny than bliss.

A good Lancaster County farm had few hedgerows, very small woodlots, lawn to the barns and lawn around the barns. No weeds. Tangle was akin to sin. Land was tillable, and therefore the land was tilled. Whatever grew was planted. And always for human use. (When I moved to Michigan and first saw the huge leaves of wild burdock, I thought, good lord, *rhubarb*! The mad Germanic notion that if it was there, it had been planted, and could probably be cooked.) In Lancaster County it would have been unseemly, a serious lapse in responsibility, to let a field, even a rocky slope, grow up in weeds. The same obsession with order and control of the landscape continues there now, even though most of the farms, because they were "scenic," are gone—sold to developers and subdivided and recut with streets of houses. Each development wanted a view of the next farm. No postcards now. But the order persists, and it follows the same tyrannical, familiar assumption: humans have the right, the obligation, to work and rearrange and "order" the natural world. The wild is a place to be tamed. It is an arrogant designation of priority—make the world over for humans. Americans, seeing landscape from the beginning as real estate, are scrupulous about dealing with it, and in prosperous areas all over the country, a fierce moral judgment falls on "waste" places, scrubland, even the vacant lots in developments, where the ragweed's got a good hold. For God's sake, do something with it! everybody says.

*   *   *

In Michigan, too, you hear remnants of a wish for control. Shreds of it. When I called my neighbor to borrow a cultipacker—the wide iron cylinder that would press down the seeds of a new native prairie patch—her question about the coneflowers and partridge pea, predictably, was, "So, will the stuff spread?" She knew it would. And her question carried the comic knowingness—the willful resignation—that comes with living where human control isn't absolute. Where weeds win.

There's a kind of what-the-hell feel to farming here. You do what you can, but the uneven ground, the clay pockets, the swamps, they're a powerful opposition. You compromise. You give in. Sometimes, you just let go. You can't delude yourself into thinking that farming is God's work, or a good man or woman's work. You lose faith in the idea of beauty as a mowed yard. You don't look at an aerial picture of your farm and think, ah, peace. Harmony.

Instead, you haul rocks. You pull out mulberry and locust trees that sprout in fields. You spray quackgrass. You say, "So, will the stuff spread?" And sometimes—when it finally makes sense—you let cracked drainage tiles stay cracked. You let the swamp be a swamp. You let the locusts sprout and take over a stony field. You stop baling hay in the wet low places where the bales came in damp and molded anyway.

I've done every one of these things, and every one is a mindful struggle. Things look worse, for a while. In a way. When I stopped farming some fields, I minded the scrappiness as much as any dead-end farmer. I thought about putting up signs—Wildlife Preserve—as an excuse, an explanation. And one of these days

I might get around to doing it. Not as an excuse anymore, but because it's the truth.

Taking the care (and it *is* care, not lassitude) to hang on to productive land and let go of the hills, the holes, the margins, some farmers have found it's even possible to make money. One farmer near here, who reduced his acreage to the good acreage, makes more money farming 140 acres than another farmer makes working 1,000 acres of unselected fields, farmed straight through, no matter what the lay of the land or the condition of the soil. But there is the "clean" farm, spread out. People say, oh, take a picture. We like these pictures, the photogenic. And in farming, that usually means as complete a control of natural conditions as possible.

A number of untamed, unadulterated beauties have survived in this country, singled out and preserved, protected. When the dream of wilderness confronts the dream of the tamed farm (Americans know how to split the soul), we've often done a good job saving our "natural wonders." But to be wonders, wild places must meet substantial requirements as dramatic (Yosemite), spectacularly bizarre (Yellowstone), grand (the Grand Canyon). Even if we've done fairly well preserving the dramatic and starkly beautiful, we'd also do well to attend to ordinary places, the nondramatic, waste, and wayside places—all habitats and sceneries.

To a farmer, it sounds simple. We must finally recognize the rights of the earth. Civil rights, human rights, women's rights, animal rights—all these move outward, expand in implication, if we keep at it, toward the planet's rights. The human compulsion, centuries old, to use land rather than inhabit it must ultimately appear barbaric, an extreme form of domination and exploitation.

We have grown rich on tyrannical ideas: the idea that the world is a resource, for human-only consumption—an idea that cleared the pines, utterly, from the state of Michigan in the last century; the idea that tillable ground should all be tilled—an idea that is clearing jungles and "bringing life" to some deserts today and causing desertification elsewhere. We have an arsenal of ideas about land use clearly as dangerous to human life on the planet as the use of nuclear arms.

And worse, we have no global plan, no serious national debate, concerning the preservation of land and landscape. In fact, in the development and use of coastlines, for instance (and lakeshores and water rights), recent court decisions persist in supporting private landowners' rights rather than the broader "public interest"—although even that interest, too often one state's or one nation's interest, can be extremely narrow. Americans who regard as sacred the right to hold private property must also ask at some point, what are the rights of the property itself? This is not simply an environmental question, although environmentalists are the ones we most often hear from (the ones who, like Earth First protesters in the Northwest lumbering districts, are labeled "eco-terrorists"). What are the rights of the earth itself? And another way of asking it is: how should humans inhabit a world not wholly human?

After all, this is a world of rock and water and air. It is elemental. It is not ours.

Our rights cannot be exclusive. Human habitation of the planet must be based on mutuality, not domination. Feminists know this territory. Some farmers do. If we care about the land, it will be necessary to redefine whole economies, not just the farm economy. A complex, solid economy could certainly grow around a policy of cooperation with natural environments. Why haven't we proposed such policies—on as grand a scale as national defense—when our own species is at stake?

We are primitives in our thinking about economies. We are babes. We believe the world is ours, like a heated house. Landscape is sold for holiday viewing. This culture and this economy promote the idea of all-terrain vehicles and disregard the idea of terrain. From an ATV, the landscape is backdrop, nothing more. It's scenery on an ad. You roar through it. That's it.

Almost every April, guys on three-wheelers rip through the beech woods at the back of this farm and cut down the stream bank. With the trillium and rue anemone, it's a pretty place to ride through, with enough rocks in the water for a scare. Once when my son and I caught up with the machines—there were three of them, stalled in some swamp willows, as far into scrub as they could drive—it was clear these were men prepared for combat: they had the camouflage waterproof gear, the helmets and goggles, buckled gloves.

We smugly dismiss the nineteenth century, its imperialisms and arrogances and abuses, as unenlightened. They were blind. We can see the arrogance and abuse. But in our own world, we see once again and only—progress. Development is one of our favorite, most blessed, words. Real estate development. Third World development. Arctic development. Development for tourism.

Instead of new housing starts (more developments, more subdivisions) as one of the measures of an economy's health, why not new reclamation starts? Why not?

* * *

After all, this is a world of rock and water and air. It is elemental. It is not ours.

What do we want to *do* with it? Because we are conscious creatures, the entire planet—the universe—has become a place for the pleasure of the human mind. And being human, we must range and speculate. We must terrify ourselves with our thinking. This is our art.

But in the dailiness of human life, in the physical world of carbon and hydrogen, oxygen and uranium, we may not range thoughtlessly or speculate endlessly to our gain. The world is not ours to use up, or blanket with our debris, or despoil.

In our time, we have made of the world an elaborate, grotesque, noxious cake—it's layered and layered with richness and artificial decoration: a global and decadent art, the mind can say. Who can eat it? Who wants to? What is the appetite that cooked it up, or could be satisfied by it? Set the globe in a great gallery, and it would be something to see.

But life is not art. Or rather, it *dare not* be. We dare not let it be, or the world will be lost. Lost to us, and lost to all artless things—the matter—of the earth. Some knowledge—of death (which is inevitable)—cannot be lived (except through art), and therefore it must be known and accepted. Other knowledges—of violence, destruction, tyranny (which is not inevitable)—cannot be lived with for long (except through art), and therefore these must be known, and rejected.

★ ★ ★

It is possible to reconceive the world. It's been done. In the past, humans changed their minds; they went from heresy to new belief. It became apparent: the Earth, the Ptolemaic center of things, was not the Galilean center. It became apparent: slavery, economically feasible, was not morally tolerable.

It must also become apparent: the physical destruction of the planet—right here, an easy mark—is a crime.

★ ★ ★

In the terror of ancient times, humans could live intimately with the natural world. Without technologies, they no doubt lived in awe. And in peril.

With our technologies, not just ones of incalculable power: earth-shattering, planet-altering; but also ones of incredible potential: earth-restoring, planet-preserving, we can rediscover an intimacy, a mutuality with the natural world, that is not primitive (even if based in part on fear), but *knowing*. It might even be possible to relearn a life of awe. And inhabit landscape without violation. With the least violation.

There will be nothing simple about living generously, coherently, and intimately with the natural world.

If I'm happy to see a few Midwest fields go wild, it's a small thing. I know. But re-viewing, re-conceiving the land we inhabit, is not a small thing. It is not nostalgia to sing the praises of vacant ground. It is not longing for the past, but an immoderate and profound desire for the future, that leads a person to say—about wasteland and wetland and any steep slope and any undeveloped shoreline—let it go. Let it be.

BURIED WATER

Water is wild, it's outlaw. It takes topsoil, it channels serious and grand canyons, it collects in wetlands and goes noplace. It stinks. It sinks, it springs. Water falls, flows, gathers, floods.

Even so, human beings want to *walk* on it. And not get their feet wet.

A miracle's a passionate, compelling story, not the usual muck and mire. And if it takes the special effects of engineering, dredging, blasting, bridgework, drainage systems, and various metal-clad machineries to work miracles in nature, well, that's good business, too.

We walk on water every day of the week in southern Michigan.

It's a classic American story—domination of the elements—and the action has been most ruthless and visionary and violent when the main players have come upon water.

Water starts out simple, very clear. But as soon as it hits the ground we claim, things get murky. And you know how much we claim: every square inch.

Look at some square inches and acres around here. Ypsilanti, for instance. A narrow, funnel-down point in the Huron River watershed, Ypsilanti has been claimed ground since the French Claim of 1809. On early maps—1825, 1874—the Huron River takes a sprawling turn at Ypsi, with marshes on both sides of the wide floodplain. A small stream flows into the river near Forest Avenue. A bit later, in 1907, on the delicate, hand-colored survey maps drawn up by Gardner S. Williams, they're still there, the marshes, that small stream.

But look around now. Walk around town, along Forest Avenue to the Eastern Michigan University campus—you won't get your feet wet. That ground where the stream should be, through the middle of campus? Dry ground, sidewalks. Water, water, is not anywhere.

Except for the Huron River itself, other visible waters—the original swamps and streams—have been disappeared from the city (as from almost every city), and it didn't take long, a few generations. Miraculous is the word people like to use when they hold Nature down for the count. The most cinematic miracles in the Bible are water ones—Moses parts the Red Sea; Jesus walks on water. But if God isn't on your side in these matters, if you can't single-handedly part a stream and cross it, well, you can bury it.

★ ★ ★

The place I work in Ypsilanti, Eastern Michigan University, stands on a slope falling away to the south bank of the Huron River.

Around EMU, the river takes that wide, formerly swampy, turn after Leforge Road, swings around to the Forest Avenue bridge and then runs more or less straight into Depot Town and through Riverside Park. The Huron River isn't buried, it isn't completely barriered, but most students at EMU don't notice the Huron River and don't think of it as part of campus. They know the river as part of the scenery in Depot Town, east of school, where Frog Island and Riverside Parks open up the water to view.

In Depot Town, you can walk on a scaffolded wooden walkway that connects the parks and crosses over the river, under the Cross Street bridge. It's the one place in town you can meander, get close to the churn of the water, hear it wash around the rocks there, and check out whatever debris has snagged in the brush and slung off to one side—a couple of tree stumps, clutches of cans in branches, a tire, some fishing lines.

On campus, though, you don't think about water flowing nearby. Everybody drives on Huron River Drive; but at Eastern, close as the river is, Huron River Drive is a *drive*. The name becomes the road, cut off from view of the water by the Riverrain apartment complex and the Eastern Plaza mini-mall.

Most students could draw you a decent map of the campus with the streets intersecting at exact angles, but the river wouldn't be there. If you said, "Draw in the river," they'd probably have to think about it. "The Huron River," you'd say. And then they might be able to backtrack the river in their minds, coming upstream from Frog Island and guessing about the turn and about what happens behind the railroad tracks, the old Peninsula Paper Company building, behind the mini-mall.

But where the river comes from—back toward Ann Arbor and Gallup Park—and where it goes—someplace after Ford Lake—that would be distant territory, unmappable.

We're explorers now in watersheds, with no signposts, few maps, or with blank territories on our maps, those drop-offs at the edges where cartographers used to draw dragons in threatening seas.

We believe we know where we are. And it's true, we have some very good maps. But, it is also true—we have no idea where we are.

We know road maps, not watershed maps. Not vegetation patterns. Not soil maps. Not buried water maps.

An address? Most of us know the street number, the ZIP code. But who knows the watershed or its number?—the digits tracking back from outlet to large rivers to streams. The Huron River Watershed: #04090005.

For many of us living and thriving in watersheds, ecosystems, and climates, those elemental systems have become deep background, lost to our thoughts and experience. The more a place is settled and built up, the more uncharted its natural features. The lay of the land, landscape, the fall and flow of water—it all disappears.

And if you decide not to bury water, you can blur it away, set it aside pretty completely. Just about every river, for our safety, is bridged and barriered—there's no drive-by viewing. Most old-style see-through pipe railings have been replaced with reinforced concrete sections, shoulder to shoulder. We drive over rivers without knowing it, without seeing their course, their width, their particular ripple. Since they are out of sight, most highway departments don't label the rivers with signs anymore. You're on a bridge, you know that, but what you're crossing—who knows? You can drive across the Midwest on interstates and not know you've crossed any river, any watershed.

Not long ago, the Cuyahoga River near Cleveland was a presence on the Ohio Turnpike. From a car on the bridge, you could look through pipe-railings, dizzyingly, at the river below and know its strange name, maybe hum a few bars of the R.E.M. tune. The river was a clear water-stripe, a twist coming out of trees, with high banks and a visible floodplain, some rocks spitting white streaks in the water. The turnpike—any driver could see—was a road built from high ground to high ground, landforms the Cuyahoga River had cut. Now, with the concrete barriers, you can't see the river, you don't know you're on a riverbank. And half the time when I drive there, even though I'm determined to hum "Cuyahoga" in tribute to the river, I miss it and drive right through, and feel too bad about it later to sing anything, retro-honorifically.

When we cross over water, we're safe. (And sorry.) It's a clear cut, a straight shot. No scenic distractions. No notion of waterways, watersheds, landforms, nature.

Out of sight. Out of mind.

Water is wild. It obeys the invisible and elemental laws of gravity, absorption, evaporation—not human laws of boundary, possession, property.

Until we get our hands and machines on it.

Water's our source, our sustenance. It bore us, it buoys us, it can bear us away.

Still, we believe, and by law we do, *own* water. In the past, we've pretty much done what we pleased with it. In the West, with water a rarity, the story's an old-time romance, a tug-of-war, an ongoing saga—with schemes to conjure up water, claim it, carry it across state lines, and marry it to dry cities, dry farms.

But in the Midwest, in southeast Michigan, where water stood around just about everywhere, such a common thing, the story has always been: ditch it, drain it, bury it, forget it.

★ ★ ★

Ditched & Buried, To Start With

When I first farmed in southern Michigan, I farmed dry ground. The fields were sandy loams that drained so fast after rains I could often work the ground the same day. The only water on the farm was a stream in the woods at the back of the property. One year a sinkhole appeared in a low spot in the hayfield and, when I checked it out, dug down into it, I pulled up a chunk of broken clay tile—terra cotta, the color of the bricks of houses around here, like a shard of a flower pot.

I had no idea. I'd never seen water collect in those fields, and I'd never thought about it. Or about Michigan soils and glaciated ground. About its watersheds and drainage. Even though I was obsessed with weather and watched for the high cirrus clouds, drawn back like hair, a sign my grandfather taught me meant a low-pressure system moving in with rain, maybe, within thirty-six hours. I farmed hay, and I had to think about rain, know when it would come, and know how long it would last. I watched the sky and I watched the TV.

But I thought rain fell and flowed downhill to the stream or percolated down into groundwater somewhere, the way it did on the farm where I grew up in Pennsylvania. There, everything either sank through fine clays into porous limestone or flowed downhill—you could see it—through the network of streams to the Conestoga Creek and on to the Susquehanna River.

But here in Michigan, with the complex and mixed soils, glacial hummocks, sands and gravels overlaying marls and clays, water collects. It stands around.

Until somebody digs a ditch, carefully sighted downslope, lays some drainage tiles, and buries the water.

On most of the farmland of Michigan now, the watershed lies underground, where water flows at a terrific pace during thaws

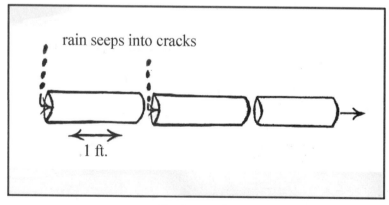

Clay drainage tiles

and rains. Buried, water is invisible, silent. You walk right over it.

All eighty acres of my farm were ditched and drained. I had no idea. There was no record of the work, no maps. But in the years I farmed those fields, the old clay-tile drainage system broke down, one tile after another. The tiles filled up with dirt or roots. Or the clay just shattered like stomped-on glass. Low spots got wet and stayed wet. Unplowed places sprouted sedges, loosestrife, and clumps of reeds.

When I plowed the bottomlands, I'd sometimes cut through the tops of strange gray-dirt chimneys that crayfish kicked up in wet ground. I could look straight down the chimney hole to water, six inches below the surface of the field. I was farming a bog. The tractor would push out a kind of waterbed ripple through the dirt.

In the end, it wasn't worth it, working that ground, and I quit farming the low spots. As more tiles broke, small wetlands formed, and the extent of the tiling was clear. This was a huge water network, a watershed underground—coursing, flowing, lying low.

I see any farm field with a different eye now. I can see what's in it, and I can guess what's under it.

In Michigan, if it looks like dry ground, you can bet somebody's buried the water.

I still don't know where all the tiles are on my farm. It's a patchwork system of arteries. A map of the old drainage system, the watershed underground, might look like this:

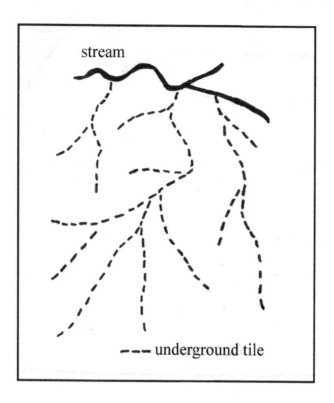

stream

--- underground tile

After walking the stream in rubber boots, I spotted several outlet points of the tile, three places in the banks of the stream. In the fields, I know the main tile lines because I've broken them now, crushed the tile, and let the water collect, as part of the federal Wetlands Reserve Program. Five huge wetlands formed—these would have been woodland swamps, in beech and maple forest, when settlers arrived here.

All that water restored, resurfaced. Out of sight!

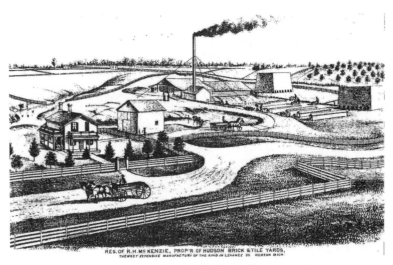

RES.OF R.H.McKENZIE, PROP'R OF HUDSON BRICK &TILE YARDS,
THE MOST EXTENSIVE MANUFACTORY OF THE KIND IN LENAWEE CO. HUDSON MICH

Hudson Tile Works (1874)

Acres of water on a dry farm.

In the decades that closed out the nineteenth century, almost all of southern Michigan's woodland swamps were cleared of trees and drained with clay tiles for farming. You can't farm in forest. And you sure can't farm in mud, wet peat, black muck, bog, cattails, and hit-and-miss mire. In many towns, along with the sawmill, the brick and tile works was the major industry.

Nobody wrote much about it, but the southern Michigan landscape in those days must have been nightmarish—trees cut, stumps uprooted, brush burning in huge heaps of flame; families in fields with shovels, or with animals harnessed to ditching blades; children hauling in and setting those tiles, one at a time, thousands and thousands of tiles in the trenched fields; then the work of shoveling dirt and covering it all up.

Burying the water.

Today, to keep a farm dry, fields are drained and redrained with plastic tiling. The process is fairly quick, surgical, with laser transits, trenching machines, and huge rolls of yellow tiling unwinding into the ditch. But the thinking and consequence is the same: ground is a surface for farming, for human use; and water must go underground.

In towns, too, like Ypsilanti, water was ditched and buried. Water had its use, but it had its limits. To run flour and saw mills, the Huron River was cut into in several places, channeling water to the mill wheels and returning it to the river. Frog Island (which is not an island at the moment) was no island to begin with, until a millrace was dredged in the 1840s, cutting a good chunk of ground away from the bank. Frog Island stood apart in the river for more than one hundred years, long after the mills were gone, until gradually, through the 1960s and 1970s, the millrace filled in with years of sediment and became a dump for concrete, construction and railroad debris.

Ypsilanti is a small point, one turn, in the Huron River watershed, but it's got it all—the love and desire of water, the loathing, the ditching, the burial.

Every water affair goes on and on. It just won't quit. Settlers drained foul swamps with clay tiles; they dredged stream banks and cut out millraces. A few years later, in a kind of wild digression, a fling in the water story, Ypsi entrepreneurs drilled bedrock wells and pumped up waters from the "Atlantis" well. Then, later, a real dissolution, a *thunk,* a hush of a close; civic leaders drained the city watershed and separated themselves from whole streams.

The *work* it took. To work these miracles. And still takes. The labors of love and of machineries. To bury water, walk on it, and walk away.

★ ★ ★

Drilled, Bottled & Bought (The Fling)

In the hierarchies of waters, surface waters are sluggish, sluttish. We don't like things easy. But, oh, the seductiveness of well water! And the deeper the water lies, the more miraculous. In Ypsi, for a few years, there was a regular crush of men, good businessmen, out to *resurrect* mineral waters from bedrock wells. They bottled and sold the water, a miracle cure—three glasses a day for whatever ills, chills, the nineteenth century gave you, whatever "disordered blood."

In 1884, on sixty-five acres of the property where EMU's central campus now stands, lived the oddly named Tubal Cain Owen. He was a nut ("a bit peculiar," the *Ypsilanti Commercial* called him), a man "of capital and push" in Ypsilanti. Tubal Cain Owen put together a small domain on this property—a wife and three children, a house on the hill in a grove of hickory trees, six and a half acres of tropical plants and lawn, with pretty Owen's Creek running through the grounds.

Tubal Cain Owen was an entrepreneur and a believer in the Book of Revelations and a great believer in scientific principles. He was a success in most of his endeavors and businesses—shipping, milling, farming—although his belief that the Bible prophesied "flying machines" led him to construct, on scientific principles, years before the Wright Brothers, an "aeroplane" in his back yard. The construction never flew, but that one failure didn't diminish T. C. Owen's faith in technology and his visions of a fantastically improved life.

Water was a curse in the lowlands for farmers, but it was also a cure one took at that time in the many popular springs and sanitariums: Saratoga Springs in New York; Hot Springs in Arkansas; Mount Clemens in Michigan. In 1883, a well drilled by the Cornwell Paper Company in Ypsilanti hit "evil smelling" water, and it didn't take many months to find backers with money to build a sanitarium, the Ypsilanti Bath House, which opened for business January 10, 1884.

Six months later, Tubal Cain Owen drilled his own, very deep, 740-foot well and struck salty water on the Forest Avenue property. A flamboyant competitor, Owen called his well "The King of Mineral Waters," and christened his product Atlantis Mineral Water. "We have no myths nor Indian legends to relate to appeal to the public's credulity," Owen pronounced in his brochure. "We sank our well on scientific principles in search of HEALING WATERS." He invited comparison with the Cornwell water, and the *Ypsilanti Commercial* wrote in June of 1884 that the Owen well water "very nearly resembles" the Cornwell water but lacks "the stench." The writer, no doubt guided by Tubal Owen in the experiment, goes on to describe "a curiosity." Rub a silver dollar with Cornwell water, and it is blackened at once. "Wash it with this [Owen well water], and it is restored to its original color."

Okay. Not quite parting the Red Sea. But Ypsi applauded. Stand-around surface water had to be ditched, but *bedrock* water—that was a moneymaker. For a while, Ypsilanti was a watering destination, with its mineral bath houses and sanitariums known "the world over."

Early photos show a derrick and a smokestack on the Owen property, but Tubal Cain Owen quickly built a brick factory around the derrick and manufactured a slew of deep-well products. He bottled Atlantis water, and Ypsilanti Mineral Water, and marketed them as "Nature's Greatest Remedy for Disordered Blood," a curative for thirty-two diseases including hay fever, hemorrhoids, pleurisy, diabetes, cuts, burns, venereal disease, bee stings, and cancer. In Owen's 1885 advertising brochure, three wineglasses a day, "no matter how nauseating it may be," were recommended; and "as an enema," he wrote, "this water has no superior."

Owen shipped mineral water in state and out of state to Chicago; he evaporated water into salts and sold that; he manufactured Atlantis Mineral Water, Salicura Soap, and Paragon Ginger Ale.

The *Detroit Herald of Commerce* claimed "the natural water has been deodorized and carbonated and makes a palatable drink of great potency in the relief of all persons afflicted with skin or blood diseases. The granulated salts obtained from the waters are highly recommended by the most prominent medical exponents for the cure of catarrh, headaches, and bilious or malarial disorders."

This water was big, flashy business. The Cornwell brothers,

backed by investor George Moorman, drilled more wells. Several Ypsilanti doctors had Cornwell-Moorman mineral water pumped to their offices. In a famous quote people still like to haul out of the archives and twist around, the *Ypsilanti Commercial* declaimed in 1884: "Ypsilanti has already come to be the centre of attraction for the halt, the lame, and the blind, the palsied, the paralytics. It is by no means a crippled city, but a city of cripples."

Disordered blood was a boon for Ypsi. The *Commercial* goes on: "The cry is every day, still they come! Let Them Come. The

Hawkins House, the Follett and the Barton, and all the other hotels and numerous nice boarding houses are full. Ere another season a mammoth hotel may be in the process of erection." An anonymous "Farmer" wrote a ballad published in the *Commercial*, with a couple of verses that tout the success and the dream of these waters:

> *If you are sad, with sickness worn,*
> *And have the headache every morn,*
> *Just come and drink a healing horn,*
> *Of Ypsilanti's water.*

> *There's forty new baths agoing,*
> *And all the healing waters flowing,*
> *Better days and health bestowing,*
> *On many a weary one . . .*

> *It's true, it has a woeful smell,*
> *But if your stomache don't rebel*
> *It's just the thing to make you well*
> *And praise up Ypsilanti.*

Owen and the others had drilled through more than two hundred feet of glacial deposits and till into bedrock, where aquifers trap Paleozoic marine waters in the sediment. Owen's mineral water was ancient sea water. Those salts were real sea salt.

Groundwater geology studies in the 1960s noted that "water obtained from bedrock in the Ypsilanti area is not potable in the ordinary sense, although small quantities are consumed." During the bath house boom years, small—and large—quantities of salty bedrock water were bought, bottled, and consumed from Ypsilanti wells.

The fling didn't last long. Around 1902, new owners of the Ypsilanti Sanitarium began admitting "many d.t. victims" whose "yells annoyed other patients and staid Ypsilantians." The social status and respectability of the bath houses declined, and in 1906, with the passage of the Pure Food and Drug Act, requiring—alas for Atlantis waters—truth in labeling, the business of miracle waters wound down. No more cures for thirty-two diseases, no more fantastical labels. The well water was bottled and marketed as simple mineral water for another decade, but then the wells were capped. There's a pipe in the ground someplace on that hillside at Eastern, probably just about under the old President's House.

★ ★ ★

Ditched & Buried, Again

With surface water, *not seeing* things is the trick, the magic, the miracle. Invisibility is better than any one-shot parting of seas. The permanent disappearance of water—so we can walk from any point A to any point B and do whatever we want with the ground between—that's the dream and deepest desire. The American dream doesn't take long to move past the manufactured pastoral and into an engineered territory of full and free enterprise.

Tubal's lawns, with Owen Creek burbling there behind the Atlantis well and manufactory, took shape in the first dream. And all that came next took shape in the next, the ongoing dream.

On an 1874 map of Ypsilanti, Owen Creek winds around and flows more or less parallel to Forest Avenue, down to the Huron River at the railroad bridge. If the creek were there now, it would run between the parking structure and the new Student Center (there's a bridge crossing an empty ditch there now), behind the I-M Sports building, down the street past Pray-Harrold, and on under the Alexander music building.

Of course, there's no stream, and no mouth of a creek downslope at the Huron River.

What there *is* at that point is a six-and-a-half-foot reinforced concrete pipe: the Owen Outlet Drain.

On early Drain Commission maps, it's the Owen *Creek* Outlet. But more recently, the maps say simply Owen Outlet. The file folder in the Washtenaw County Drain Commission office is labeled Owen Outlet. No creek no more. It's ditched, piped, buried. Done for.

Like the wetlands that farmers drained and tiled underground, many streams, and some rivers—the Grand River in Jackson, for instance—were buried in Michigan towns, to facilitate city construction, and to control the stormwater and flooding caused by that construction.

Gulf air might build up in weather systems and dump on EMU, but the rain that falls no longer follows the watershed topography into Owen Creek; it's channeled underground in the Owen Outlet Drain. Grates in parking lots and cuts in sidewalks collect water into a network of buried streams and tributaries. But in spite of the extent of storm drains—or because of their miraculous invisibility—buried water remains somehow mysterious, a hidden knowledge. Who knows where the water goes? A few engineers?

Not many people in Ypsilanti know. The map of the storm drains on file in the Ypsilanti Office of Public Works was drawn in 1960! It's a beautiful yellowed map, frayed all around, with a few penciled additions.

"They told me to guard this with my life," the woman said, who set it out on a table. "It's the only map we've got."

"Nothing more recent? How do they know where to fix these things?"

"Well, some of these guys have worked here a long time."

Buried water's a mystery. Cultish. Out of sight. Secret,

Owen Drain Outlet, Ypsilanti

forgotten, hushed-up. Did somebody say *repressed*?

At EMU, drawings of the campus' storm drains, including the Owen Outlet Drain right down the middle, date from 1970, before the Alexander music building was built on top of the drain. If you put your ear to the ground during spring thaws, could you catch a murmur? A hum?

When talking to people at Eastern, only Dan Klenczar, then project manager for EMU's physical plant, knew about Owen Creek. He was the one person I found who knew that there had been a creek on campus, and that the creek had been put into a drain. He wasn't sure when. The County Drain Commission knew the date of construction: 1929. Almost eighty years ago. Nobody knew why, although "construction, probably," "expansion," was a good guess.

In some places, eighty years would be recent history. But here, and in America generally, landscape memory is short-term memory. And when we bury water, the first thing we want to do is forget it.

Still, sometimes old names carry watershed history. Check out a map of the Huron River watershed. Ypsilanti's the base of a bottleneck. The watershed narrows there between two hills, and the river starts to pour itself out toward Lake Erie. At this point in the watershed, you can literally see from one side of it to the other. From the hill with the Ypsi water tower, on Summit Street, you can sight across to the other side—the old Highland Cemetery and beyond that, Prospect Road. The names recall what the settlers and early surveyors knew. What we've forgotten.

On the other sides of these hills, water flows away through other watersheds, finds different rivers and different outlets to Lake Erie.

Without a map, who would know? We count on the names: Summit, Highland. But what if the name shifts—Owen Creek Outlet to Owen Outlet? The creek is gone from the maps, gone from the name. Out of sight. Out of mind.

When it comes to water, we've lost our senses. We bury water. We forget we bury it.

Who buried Owen Creek?

It wasn't Tubal Cain Owen. The stream fit into his lavish and sweeping design of the property. And nobody then knew the scientific principles for the construction of pipe big enough to hold the stream. By the time Tubal Owen died in 1913, the State of Michigan was anxious to purchase the land, to allow the Normal School, now EMU, to expand. The buildings of the sanitarium had been sold for residences, although water from the Atlantis well was still pumped by Owen's son and shipped in bulk to Chicago and Boston. EMU's president Charles McKenny argued that the Atlantis well and facilities, though in competent hands with Owen's son, could become "a serious menace" if passed on to other interests.

The Owen family resisted sale of the property, battled the state in court, but lost; and not long after World War I, the Normal School gained title to all the property, including the well, the stream, the homestead, the old factory buildings. The place would be cleared to make way for, one after the other, buildings on the expanding campus: Roosevelt, Jones-Goddard, Pray-Harrold, the I-M building, Downing, Quirk, Alexander.

About the time of the Owen buyout, in the spring of 1918, another event may have contributed to the fate of Owen Creek. With heavy snows and a sudden thaw that year, the Huron River flooded through Ypsilanti, breaking both Superior and Peninsular Dams, and inundating the low-lying parts of town. No doubt Owen Creek flooded as well. As a consequence, the Washtenaw County Drain Commission began a series of storm drain constructions along Owen Creek, first a short drain near the source, in 1922; and an extension drain in 1925. The main section of Owen Creek remained intact, though, flowing all the way through the campus and into the Huron River.

In the end, it wasn't fear of flood so much as love of power— the power of light—that buried Owen Creek.

In May of 1929, Clayton Deare, the Washtenaw County Drain

commissioner, wrote a letter to Michigan's governor, Fred W. Green, asking for state funds to cover drainwork on the state's property at the Normal School. He wrote, "As you formerly lived in Ypsilanti, you will probably recall the old creek that empties into the Huron River near the Lake Shore Freight Depot [Forest Avenue]. A petition has been filed with me to inclose this drain. It passes in between houses and continues on up through the Normal property. I have talked with President McKenny who stated they would like the drain inclosed on part of the State Property for the reason that they intend to build a new power house on the spot where this creek now lies."

A power plant it would be. And continues to be. On the spot where the creek now lies buried.

The Washtenaw County Drain Commission put out a call for bids on the Owen Creek Outlet Drain, specifying precast reinforced concrete pipe, six feet six inches in diameter, for a drain that would follow "the general course of a creek, at some places being in the present creek bed and in other places in a new course."

Water and power. Power and light. To have light, we ditch and drain a creek. Or dam a river.

We walk on water. We run on water.

We have the power to do what we want.

On the slopes and streets of Ypsilanti, water falls and it disappears. It goes underground. We forget all about the sprawl and the spray of water, its name, its course through the watershed, its stink or its skid over rocks, the body-roll turns, and then we forget we've forgotten.

The maps of water and watersheds are ancient—as rare, as recondite, as maps of the psyche. Memory fails, and fast.

It may be a miracle to walk on water. Make that wishy-washy stream look like ground. Tame it. Shut it up, shut it down. Forget it.

But, Lethe, remember, that river of forgetfulness, we cross over into hell. And then if we bury the river! There it is—we forget we've forgotten. The place may be tillable, a good site for a power plant, but how will we know where we are when the water's buried and there's no map?

What kind of miracle is it, if we don't know we're walking on water?

MALINGER, MEANDER,
IN PERPETUITY (A CREED)

malinger—v. Like dropouts, ditching the grind, AWOL, we
malinger and offer plentiful excuses: *we are in mourning in*
perpetuity, we were wheezing when we should have been kissing, we
couldn't drink from that pool, etc. When the sun finally appears just
behind the fog, blurred and lopsided, we detour, we strike out,
meander. Get lost. This is the way to see the world, offside, out
of bounds. The only way to commiserate with shale, say, or the
seams of the ocean floor opening, molten. Or with the living, the
liverworts flattened on rocks in the stream, pigs, the mink that got
the wood duck, the top of the top of the canopy, parasitic worms.
The world rolls, shifts from one point of view to another, the
planet is replete. Enough is exactly enough. Sloughing the scales
from our eyes, outlaw for a change, we see dust, and the broad
colors of decay, the purples especially iridescent, the mother of
pearl. Wandering along a wall, any wall, we see past it. And into

it, the shells, the corals. That's about it for power. A sufficiency, a
kinship with whatever is not everlasting.

meander—n. also v. Not in a line, but wandering all over the
place, all over the page, too. Just get in the swing. The flow
of water meanders, it has to, something gives, nothing in the
world is uniform in resistance or fall. When the interference is
slight, almost imperceptible, not granite slabs, downed trees,
or dams, not machineries, nothing like that, but your ordinary
less-consolidated soil, or porosity of stone, something that caves
bit by bit, well, there it is, the water washes in, takes its turn,
washes out, a meander. There is no motion more mild, a turn
into and a turn away. Nobody slaps anybody. Both sides give in,
pacific, accommodating. Unlike the well-known hard-ass linear, a
headlong and cataclysmic fall down a two-thousand-foot rock-face,
a drama that brings in a crowd—that's another story. It begins, it
ends. It happens in one place, you can open up a folding chair and
sit there and watch. But meandering goes on and on, you have
to follow it to see what happens next, you never know. You're in
the midst, not a spectator, you're ambling, wandering. Meanders
swing around, sometimes they circle back, cut themselves off, but
keep right on going. Float a meandering stream, build a makeshift
raft, that's fine, you spin leisurely, you dangle your feet in the
water, there's no danger, just some confusion when one turn
looks like another, but meandering isn't a plan and you couldn't
say where it all started or where it will play out. It's not a bad plot
line, a meander, if plot is—as it is—ground.

perpetuity, in—phrase. Endless, forever and ever, *in perpetuity*
can only be said, with any conviction, of the cosmological.
And there could be question there. Not much goes on and on
eternally, thank god. Not writing, not Rocky Mountains, not
real estate, not names, not agriculture, not artificial intelligence
or any intelligence, not good works, not one species or another.
Mites, dinosaurs, solar systems come and go. The procession, the
overthrows and obliterations, deny perpetuity. Still, on a piece of
paper, the seventy-acre easement on my farm goes on and on, *in
perpetuity*. No structures, no farming, no commercial enterprises
of any kind, no draining wetlands, no nothing, *in perpetuity*. On
paper (what else is flimsier, less likely to last, in our hands?) they
are words to say to yourself, chant them, a consolation when there
is no consolation, here is something for nothing, out of our hands,
rock is rock in perpetuity, wetland wetland, dust dust, however
long there is dust, and whatever follows follows, out of our hands,
long gone, we're meandering now, that's the way to go, year after
year, waving ourselves good-bye.

malinger, meander, in perpetuity *(A Creed)*   137

THIS STREAM, THAT STREAM

The stream on my farm, St. Joseph Creek, meanders through rich floodplains with cardinal flower, sweet Cicely, stands of old beeches, and understory colonies of pawpaws, leafed for the tropics. Pawpaws bear a custardy fruit, the Michigan banana, in early October. The fruits hang singly or in pairs high up in the trees. They're not easy to see, with skin that's leaf-green, and smooth as your own skin. But when you spot a few up there, picking's a party. In that tangle of prickly ash and wild grapevines, you designate somebody as spotter, and then shake the slender pawpaw trunks. If the fruits are ripe, they fall with a *thub, thub,* and the spotter nabs them.

The flesh is so soft, we carry spoons to the woods, slice open the pawpaws, and scoop the bright yellow mush right there. It tastes dreamy—aromatic and floral, musk at the edges—as if it had dropped somehow from a jungle canopy or cloud forest, a taste of Trinidad.

But this is Michigan, and the St. Joseph Creek flows east to join Bean Creek, which heads south to Ohio, into the Maumee River for the last wide stretch to Lake Erie.

★　★　★

They say in Michigan you're never more than six miles from a lake or stream. Where I live in southern Michigan, you're also never more than six miles from a Confined Animal Feeding Operation (CAFO)—a livestock factory. We've got a dozen of these factory operations, some with hogs, but most of them are dairies, with thousands of cows confined year-round.

Fifteen thousand cows—you can picture it—produce a huge waste stream. And it's *literally* a stream. The animal waste is liquefied with clean groundwater and pumped to lagoons, holding pits the size of small lakes.

Here, I'm never more than six miles from a manure lagoon.

★　★　★

Not far upstream from my farm, about five miles away, a slope in the woods is the drainage divide that shapes the headwaters of the St. Joseph Creek. Several springs contribute clear water along the way. The creek winds through woods, falls through cobble and gravel runs, cuts a wider swath through acres of marshland, and then drops and takes S-turns through the beech woods of my farm, before opening into flatter farmland downstream.

Nobody's built a livestock factory upstream from my farm, or dumped liquid manure on fields there—yet. Just downstream, though, it's another story.

★ ★ ★

Since the first livestock factory was constructed here in 1997, the operations have had pollution problems. The pit-and-pump system contaminates water coming and going, from the barns to the pits, and that's just the beginning of it. To see the worst, when liquid manure is sprayed on fields, you have to get down in the ditches. You have to know what's happening underground.

Buried under farm fields throughout the Great Lakes Basin is a complex network, estimated at *millions of miles,* of drainage pipes called tiles. Agricultural tiles drain away rainwater; that's their purpose. But they also carry farmland pollution directly to streams. Out-of-sight, subsurface drainage is agriculture's dirty secret.

Some field tile systems are new, some are a hundred years old, with fixes from every generation along the way. Tile outlets to streams may be far off a farmer's property.

In many soils, wormholes—more than you'd think—and macropores, large cracks in glacial soils, are direct pathways, like straws, to subsurface tiles. Liquids can move from the surface to tile pipelines at incredible speed, sometimes in just a few minutes.

When a tanker sprays black manure water on a field, check the lay of the land, the drainage, and go downstream to a tile outlet, and wait. Watch that pipe.

★ ★ ★

St. Joseph Creek is a natural stream. In Michigan that means you can't, legally, dredge it, and you can't bury it. But many streams flowing through flatter land—the ancient lake-beds of Michigan and Ohio, the Black Swamp south of here—were designated as "drains," in the late 1800s, giving farmers the right to channelize them and to dredge them. Field tiles jut out into these ditches at regular intervals, every thirty feet, or every sixty feet, pouring the field-drained water into the drains. Drains which are streams. Tributaries of the Great Lakes.

* * *

Every few weeks for two years, I've been out in the ditches, monitoring the water. I drop the probe of a dissolved oxygen meter into the stream for a digital readout; Kathy in her babushka, or another volunteer, extends a long-handled grabber with a sterile bottle attached and scoops out a water sample for bacterial testing.

Like many others around here, I got into watershed watchdogging after a massive manure "spill" in March 2000. Millions of gallons of polluted water flowed into Lake Hudson, a nearby state park. I joined a grassroots organization in Hudson—Environmentally Concerned Citizens of South Central Michigan (ECCSCM)—that advocates for responsible agriculture. We set up a monitoring project, and for two years we've been sending our test results to the Michigan Department of Environmental Quality, to County Health Departments, and to any legislator who'll listen.

Many politicians would rather not hear about pollution from

farms, from factory farms. Industrial agriculture proclaims its conservation ethic; livestock producers say, "Heavens, we wouldn't pollute our own land, our own livelihood." But if farmers don't pollute, the liquid waste sure does.

People around here keep an eye on the ditches and call up Lynn, another ECCSCM volunteer, when they smell foul water or see discoloration in the streams. Lynn drives her pick-up around the livestock facilities several times a week, checking drains and streams. When pollution flows, we try to get *E. coli* samples before everything's gone downstream. We take photos of brown water, green water, black water—we've seen it all.

★ ★ ★

In April, the American brook lampreys, like small silvery eels, spawn in the rippling place where I set a bench, for listening to the best stream sounds. Out of the gravel substrate, the lampreys surface after half a dozen years in larval form. Living just a few days as adults, they spawn in these stretches of cold water, twining in clusters and twisting as their bodies exert tremendous effort, latching onto pebbles with their sucker mouths, moving the stones to make a nest.

Freshwater mussels live in the sandy holes and cobble runs of St. Joseph Creek—slippershell, fatmucket, round pigtoe, rainbow, pocketbook, fingernail, spike, giant floater. I find empty shells on the shore where raccoons have feasted. If I put on my rubber boots and walk up the stream, I can see the live ones, with shells slightly open, siphoning and filtering water.

Mussels, native freshwater lampreys, mottled sculpin, redside dace, water pennies, stoneflies—they all need cool, clear, and unpolluted water to survive. The mussels can't relocate if the water is fouled. The American brook lampreys live where they hatch, only those graveled and rippling neighborhoods.

★ ★ ★

In cities, stormwater usually drains underground through one set of pipes, and human waste flows through a separate sewer system to treatment plants. "Combined" sewers have been a serious municipal problem in the past, causing flooding and overflows contaminated with human waste.

But agriculture has largely ignored its own combined—and continuous—sewer contributions to waterways.

About 80 percent of the fields in this part of Michigan are drained with subsurface tiles. The first thing the big dairies did here was bring in equipment to retile the fields where liquid manure would be sprayed. These new grids drain water and wastes even more quickly, more completely, than the old tiles.

Some manure haulers plug tile outlets with inflatable devices to halt the flow; but sooner or later, it rains or thaws, and they have to pull the plugs. The pollution flows. Neighbors call it plug-and-play.

★ ★ ★

Toad Creek is a stream through farmland, another tributary of Bean Creek. It's in flatter territory, a designated drain, and the official drain map calls it Blanchard Joint County Drain. When I first saw Toad Creek a few years ago, it was a clear stream, with plenty of toads, fish, and mussel shells strewn on the sandbars. Shrubs along the banks gave some shade and shelter for wildlife. Since then, two big dairy operations have been spraying liquid manure in the vicinity. Drainwork was done, removing the vegetation—"cleaning" the drain. And in March 2003, thousands of gallons of liquid manure were sprayed on a field, on frozen ground. Within hours, the contaminated waste ran off the field, down a drain-inlet, and reached Toad Creek through the underground plumbing.

The liquid manure poured out a pipe onto the ice of snow-dusted, frozen Toad Creek. It flowed on the ice for days, a black stream.

Through May and June, Toad Creek was often grayish, with scum floating along the banks. The dissolved oxygen levels plummeted and bottomed out at fish-kill levels. I haven't seen a mussel in the stream, or a shell on the streambank, all summer.

★ ★ ★

The pollution from CAFOs has raised real alarms—finally—about tile drainage. Instead of trying to stop liquid animal waste from flowing, with stopgap measures in the field, livestock production should change the whole waste system and dry up. There are plenty of industrial models for dry systems of waste handling. It's not a huge leap; agriculture doesn't have to reinvent the wheel.

Or, if CAFOs won't stop liquefying manure, then these livestock factories should be required to treat their sewage, just as other industries and municipalities must do.

Better yet, livestock operations could downsize to farm size.

Get lean and green. It can't happen soon enough. Managed rotational grazing, with winter composting, brings benefits to agricultural communities, instead of the harms and health risks that come with CAFOs. Set up the solar electric fences! Plant some green pastures. Bring the cows home.

★ ★ ★

In the shade of sycamores, beeches, maples, nannyberry, witch hazel, ironwood, and buttonbush, the St. Joseph Creek stays cool. The roots of the trees hold the soil. The leaves fall in the water, and caddisfly larvae attach there, or they stick scraps of leaf on their bodies and crawl the streambed in camouflage. Stoneflies and dragonfly larvae congregate under rocks, part of the complex web of living and dying and feasting.

When I'm in the water, that's what I think—I'm in the water. I'm by and of and for the water. It's my democracy.

Homeland and habitat, every watershed is worth protecting. Worth celebrating. Water's in our blood, it's our lifeline, and it binds us. To stoneflies and stones, to skunk cabbage and clams, to rotting leaves and cooking cake.

Every spring, we have a party at the stream when the brook lampreys spawn. Cut the cake. We celebrate what they have, while they have it.

SKINHEAD AGRICULTURE

It's January and I'm driving on Dillon Highway, a dirt road three miles from my house. Well, the road's not all dirt. The first quarter mile is paved, with one million dollars in taxpayer-subsidized cash, to the entrance drive of a three-thousand-cow dairy operation, where the pavement stops. I'm down the road from there, on the dirt end, driving between two harvested corn fields that stretch more than half a mile east and west to distant tree lines. The fields were fall-plowed, so they're bare dirt now, the surface clumped and uneven, and worked so close to the road that in some places you can't tell the edge of the road from the field. No buffer—hardly a weed, not a rooted shrub, no vegetated right-of-way. Dirt to dirt.

We've had a wet fall, a globally warm, weird winter with no snow yet, no deep freezes. After rain last week, the ground is saturated. Small lakes have formed in low spots, and in many places, rivulets carve a flowpath out of the field. Muddy water runs down the road.

Everything's mud brown—the dirt road, the bare fields, the tree lines at the far horizon. It's winter, true. But I've driven this road in May, and the fields are bare and brown then as well, the ragged clumps worn down into undulations. Sometimes the fields are black in spring, with liquid manure sprayed on the ground. Even in June, these fields are still bare, the dirt scraped over, pressed down, and imprinted with the parallel tracks, back and forth, of the corn planter. Under the ground, the late-planted corn is just swelling, beginning to put down a root and send up a first shoot.

If you're lucky enough to recall farmland as lush green fields, greener pastures, and dense thick-leaved woodlots, take another look. And look again, any day, October to June.

If you're anywhere in the Corn Belt, or Soybean or Cattle Belt—and that means most American farmland from the Appalachians to the Rockies—you'll see fields laid bare as much as nine months of the year. The annual row crops are either not yet planted or already cut or chopped. The soils on fields lie denuded and exposed.

The ground looks shaved, scarred, or skinned. And it is.

Bare earth—ground without cover, without rooted plants holding soils and building them—is not a natural condition in a place like this. In its natural state, all this ground would be covered with perennial plants, rooted year-round, leafing and dying in sequence, the soil continually nourished, cooled, protected.

A field of bare soil heats up like a hot plate. Bare earth is soon enough scorched earth. The surface crusts, then cracks as moisture escapes. If rain falls, it flows off crusted areas, it gouges and erodes, and mud, which of course is soil, runs off. If it's dry and the wind blows, dust, which of course is soil, blows away.

Around here, this shorn landscape—call it skinhead agriculture—is linked with Confined Animal Feeding Operations like the dairy on Dillon. That linkage has a further scorching effect: industrial livestock production, with its manures stored in open-air lagoons and then sprayed on fields, contributes significantly to global warming, sending up huge amounts of greenhouse gases, as a recent United Nations Food and Agriculture Organization report, *Livestock's Long Shadow*, noted— 35 percent of all methane emissions, 65 percent of nitrous oxide, and 64 percent of ammonia.

When you add up the acres worldwide, it's a shock—nearly 30 percent of Earth's entire land surface is used for livestock production and livestock feed production (especially corn). That's a global-size hotspot in global warming.

Skinhead agriculture, monoculture *in extremis,* is a relatively new phenomenon, and hand in hand with CAFOs, it's growing as if there's no tomorrow. Except for alfalfa fields at dairy CAFOs, crop fields are dedicated to the annual grains—corn, corn, and corn, beans once in a while. All other vegetation is competition, sprayed with herbicides, or uprooted with the blunt force regimen of plowing and disking. With CAFOs, there's no longer "green manure," the traditional use of green cover crops like rye or timothy, seeded after harvest and turned under in spring as fertilizer. Low-impact, no-till farming, too, is gone. No-till, a conservation practice with no plowing, no turning of soils, was promoted in a major way after the Clean Water Act of 1972. But CAFOs can't manage it, since no-till fields develop cracks, and liquid manure can seep through those cracks and pollute water sources. Not only that, no-till is no good for CAFOs because the stench from manure application can be cut only by turning the whole mess under, mixing it up with dirt. And that's *tilling,* not no-till. The ground is churned up, turned over, exposed to winds, rains, and snowmelt—all the weathering that takes soil off the field by water or by air.

So here in southern Michigan, and elsewhere in the Corn Belt, it's common agricultural practice to let fields lie for months unnaturally bare. Even in midsummer, with row crops growing, between the rows and rows of corn lie rows and rows of bare earth.

There are brutal consequences to skinhead agriculture and its scraping away of green cover and rootedness. From these bare fields each year flow and blow more than a billion tons of sediment and the pollutants bound to it, a degradation of America's soils, air, and water.

It's no exaggeration to call bare-earth farming brutal. This is an agriculture divorced from good husbandry, ignorant of ecological systems, which lie beyond its bottom line. It abuses the soils and waters it depends upon and (still) claims to protect and conserve.

With no winter cover crops, few perennial crops or pastures, and fewer hedgerows and green stream buffers, that's a nine-month siege of runoff from skinhead fields.

Stop the mudness! is the slogan for a Great Lakes Commission sediment-reduction campaign. The *mudness*—sediment—is the main pollutant in the Great Lakes, as it is in almost all of America's watersheds, with agriculture the major polluter.

Stop the mudness! offers this advice to farmers who see runoff from their fields:

Change your tillage practices—don't plow so much, especially in the fall.

Add a grass or legume to your crop rotation—keep more roots holding more soil.

Install buffer strips—plant trees and perennial grasses to filter sediment before it enters streams.

The simple counsel is common sense. It should go without saying. But it no longer goes without saying. With fencerow-to-fencerow overplanting of annual grains, common sense no longer holds. It's news. Plant buffer strips! And the news has to be delivered again and again—given the monomania for monoculture of industrialized agriculture, or "catastrophic" agriculture as Richard Manning calls it in *Against the Grain*.

The catastrophe is manifest, and global. Which doesn't mean it's easy to see.

People drive by corn fields in summer, or they drive by miles of bare fields in fall, winter, and spring—and they think, ah, the scenery looks so fine, natural. But it's not natural. Look again. Bare earth. Scorched earth. It's *mudness*.

If you get out of the car on Dillon Highway, though, you begin to see what's happening.

I come here often to check out Durfee Creek, to see how the water looks at the road culvert, what color it is, how murky it is. Sometimes I check the dissolved oxygen levels with a handheld meter, its probe on a twelve-foot cable. From the culvert, I drop the probe into the water, and throughout five years of monitoring, almost always the dissolved oxygen reads at dead-zone, fish-kill level.

I saw a frog at the edge of the water once. Not much else. About a year ago, somebody threw a car door and other car parts into the water. Then a stack of boards. The junk is still there, mired in the mud. And one morning this fall, a black trash bag was half submerged, oozing unidentifiable gunk (meat? cow parts?), the mass of torn plastic writhing with white maggots. The maggots roiled in swirls, like puffed, living rice. When I bent over the culvert to drop in the DO probe, I could hear a buzz, as loud as bees, the thousands of maggot bodies swarming together in their feeding.

Durfee Creek is a dump, Dillon Highway a wreck. There's a tipping point in landscapes, when the chopping of roadside shrubs, clearcutting of hedgerows, eroding fields, illegal dumping, manure spraying, and air polluting add up, and the total is: *trashed*. This vandalism of landscape, like broken windows and stripped cars on a city street, can tip a place from a livable habitat to something not only ugly but dangerous.

One sort of damage begets another. Abuse quickly becomes institutional. Drain Commissions are called in by bare-earth farmers when silt builds up in the streams. Skinhead agriculture insists on a fast flow, a quick exit of muddy water downstream. Where I live, the County Drain Commission "cleans" a stream by scraping vegetation from the stream banks, uprooting trees, and then dredging the silt from the stream channel. Like any machine-fix in nature, it doesn't last long. With no roots to hold them, the denuded banks cave in, the skinned slopes erode—so there's always reason to return. And dredge again.

Stop the mudness!

The County Road Commission brought its machinery to this dirt stretch of Dillon recently, widening it for the massive CAFO equipment, tankers, manure haulers. They graded an extra three feet, pushing dirt and gravel onto the culvert, some of it falling into Durfee Creek. More mud in the water, dirt on the maggots, dirt on the car parts down there. Dirt on dirt.

Degraded streams are so much the norm around here, and bare-earth fields so commonplace now, it's possible to forget how recently—just a few generations ago—farmers tore off the green cover, plowed up pastures, quit using cover crops, and turned the landscape into row after row of annuals planted in bare dirt fields.

Bare dirt can look natural if you see it all the time. It's possible to forget the most basic conservation values, the simplest natural laws and ecological relationships. We lose the evidence of biodiversity, of soil health and replenishment, of natural cycles. There's no sign of it, no clue.

Just a few generations ago, this rural landscape would have been what most people still think of as countryside, a landscape with many small dairies, perennial pastures, and cows grazing outside eight or nine months of the year. Fields would have been smaller, a patchwork with green hedgerow boundaries. U.S. Department of Agriculture (USDA) aerial photos from the 1930s show large wetlands in this area, now drained with subsurface tiles or drained by channelized ditches and streams. Woodlots were much larger then, trees still a value to farm families for food—maple syrup, nuts—as well as for lumber and fuel. Farmers would have raised various row crops on a multiyear rotation, and planted cover crops, keeping more of the fields green more months of the year.

As recently as the 1980s, with the Conservation Reserve Program (CRP), many marginal fields on slopes and in wet lowlands were set aside in mixed perennial cover, the soil secured. But not for long. By the late 1990s, we were hit with the CAFO onslaught and its premium on fields with annual row crops—where liquid manure could be sprayed. In a flash, the green CRP ground was plowed under, gone. In its place, mudness. The übercultivation of the corn/corn rotation.

We lost not only the rooted fields but rooted greenways along streams as well. If you could zoom to Durfee Creek just those few generations ago, you'd probably have found forested buffers, large trees like sycamores with lateral roots gripping the soils, and understory native shrubs at the stream edge, nannyberry or wild cranberry with branches arching over the water, shading and cooling it.

Even now, you can find a few rare, undisturbed streams near here and see for yourself how Nature avoids mudness, with sequences of species through time and layers of species through space: ground covers, wildflowers, shrubs, and trees. With leaf litter always collecting, it's hard to find even a square inch of bare soil.

If you watched through the seasons at a natural stream, a snapshot a week, you'd see a time-lapse of upshoots and downfalls, hundreds of comings and goings of plants, biodiversity in action, innumerable plants cycling through bloom, fruit, seed, and decay—an endless cycle that builds and fertilizes soil. And just blows the mind with beauty.

Along the stream at my farm near the big S-curve, I've seen harbinger-of-spring in March, and in April, trillium, hepatica, bloodroot, trout lily, followed by the whole splurge of wildflowers in May—anemone, columbine, Virginia waterleaf, sweet Cicely, baneberry, wild geranium, violets, foamflower, and on and on. June and July are almost as jammed. And from August to October, when most people don't expect wildflowers, there's an even larger crowd, the taller species grabbing at sunlight—great lobelia, wing-stem, coneflowers, asters, tall sunflower. And that's just the *first* story, the floor of the floodplain.

This ecological mesh—plants and soils, microbiota and macro-invertebrates, animals, vegetables, and minerals—protects water and air, protects us. It's the web of life, a closed-loop food chain with no waste. A complex system—but so simple—founded on greenery and rootedness.

The year-round cover of multilayered perennials offers innumerable benefits, from the wealth of the seedbank, the biodiversity of species with leaves photosynthesizing, collecting solar energy, cycling nutrients, sequestering carbon, recharging the soil with organic matter, creating rich pores and crumb structures, a spongy dark soil that absorbs and stores water. Without this sponge on the sides of stream and tree roots holding the soil, water has little protection.

You clear the stream and the stream bank, it's mudness.

Which is why even USDA guidelines suggest vegetated buffers of thirty-five feet, minimum. For greater protection, three hundred feet of forested buffers are recommended. But since skinhead agriculture calculates value only as bushels-per-acre, buffer width doesn't figure in.

Farmers keep chipping away at stream edges, for a slice of an acre here, another slice there.

Along Durfee Creek at Dillon Highway, there's a foot or two of buffer, no more, and the greenery is just grass, shallow-rooted, the least hold on the least soil. Maybe an elderberry shrub here or there. That's it for biodiversity, that's it for water protection. There's no sequence, no cycle, no sign of the web of life.

What we've got is the mudness. The madness. The loss of our senses. We wash away not only the land that sustains us, but we've taken away the simplest sensuous pleasures in life—shade, clean water, fishing, wading.

Walking barefoot, splashing in a cool stream, should be child's play everywhere. In Durfee Creek you'd need neoprene waders you could hose down, and rubber gloves.

Things look bad enough close up on Dillon Highway. But the scale of brutality goes beyond the road, the field, the stream, the body, beyond the local and immediate to the long-term and global. When you step back and see the big picture, the larger domain of skinhead agriculture, the extent of its damage, you can see what loss of these simple things—greenery and cover—means for the planet.

These days, with satellite and digital imagery, we can literally see on that scale, see the sediment from the Maumee River (from here) pouring into Lake Erie, the swath of mud from the Mississippi (from a dozen Corn Belt states), splaying out into the Gulf of Mexico.

From satellite photos, the USDA compiles a "greenness index"—year-round, two-week summaries of the "Vegetation Condition," images that measure the chlorophyll activity of forests, grasslands, and crops (see http://www.nass.usda. gov/research/avhrr/avhrrmnu.htm). Shades of green depict "vegetation vigor," with the darkest shades indicating the most vigorous activity. Yellows and browns indicate the other end of the photosynthesis scale, low to no activity. On these images, you could draw the outlines of the Corn Belt, still yellow/brown as late as mid-June—the leaves hardly out of the ground. The dark green "vigor" does not appear until mid-July. There's that flash of green summer growth, then the images turn yellow and brown again by mid-September.

That's the visible measure of skinhead agriculture, its lack of cover, lack of protection, nine months—three seasons!—the bare earth of the Corn Belt.

On the other hand, places with perennial cover and perennial crops—pastureland, forests, even suburban lawns, mixed farming areas—are dark green, drawing solar energy, sequestering carbon, protecting watersheds throughout nine months, spring, summer, and fall.

Skinhead agriculture is not carved in stone. In the history of agriculture, the industrial-production model of wide-swath monocultures is a young model, just flexing its muscles, relying on brute force and the elimination of competition. Its brutal ecological damage is not ordained by God or by decree (only by farm bills, corporate ag funding of research, and government subsidies).

Agricultural policy and practice must return to its roots, to rootedness. It's a global issue now, a climate-change issue, as well as a moral issue—a choice we have to make, and soon, about how to sustain and nurture Nature, and human life. How do we want to treat our neighbors? How can we do the least harm? How can we care for our living community, the earth?

Skinhead agriculture does not ask these questions. It is amoral at best, ignoring anything beyond the bottom line and field boundaries. It is immoral much more often—knowingly polluting air, knowingly polluting streams.

The Farm Bureau and other agribusiness lobbyists want to redefine field runoff as "agricultural stormwater," a phrase that cleans it right up and hides the truth, which is that muddy water is sediment, a pollutant.

If mud doesn't sound lethal as pollution, it's probably because mud is such good fun on land, great for stomping and squishing between your toes, one of those prime recollections of pleasure in childhood. But mud in water is, in fact, suspended solids, which can clog the gills of fish, suffocating them. It buries fish nesting areas, buries their eggs. Mud can kill freshwater mussels, those stuck-in-the-substrate clams that siphon and cleanse the water in the course of their feeding. Mud is expensive to clean up, to filter out of drinking water. The phosphorus and other nutrients (often from liquid manure) that sediment carries with it can overfertilize algae in a bloom-and-bust cycle, crashing the oxygen levels in the water, creating dead zones.

In this CAFO watershed, even a flowing stream like Durfee Creek has become a dead zone. And downstream, wherever downstream is from bare-earth fields, Lake Erie or the Gulf of Mexico, large, catastrophic dead zones have formed as well, killing fish, clams, and shrimp, killing fishing economies.

What we need is a more seasoned and natural model for farm policy, an efficient and productive model—and oh, why not more sensuous! and attractive, even self-supporting!

If Nature's efficiencies, bounty, and protections were the model, instead of the industrial-production model, what would USDA policies support? How would the landscape look?

Well, much much greener. The highest priority would be— *cover the ground.* That's the only long-term security—for our food, for the farmer, for ecosystems, for air and water—so why not take global protection seriously and green the ground?

Cover the ground. Agricultural policy should *promote*—and if we're going to subsidize, subsidize *this*—perennial crops, biodiversity, the least disturbance of soils, grass-based livestock production, reforestation of hedgerows and woodlots and road rights-of-way, restoration of fragile natural areas that protect air and water. Ag energy policy should promote perennial crops like switchgrass for ethanol production (not the bare-earth annual corn). And agricultural policy should *prohibit* clearing or cultivation of riparian zones and floodplains, clearcutting of woodlots, dredging of streams.

Perennial crops are no dreamy dream. Independent agricultural research (outside the land-grant, corporate-funded research) at places like The Land Institute in Salina, Kansas, offers a glimpse of the future with groundbreaking, ground-saving perennial seed crops. For instance, David Van Tassel, a plant breeder at The Land Institute, is working to perennialize the annual crop sunflower, that familiar plant with one big seed head, drooping near harvest, a plant row-cropped now like corn with the same bare-earth brutalities as corn. A perennial sunflower selected from crosses with wild ones, favoring multiple heads and large seeds, would emerge much earlier in the spring than an annual could be planted, a boon to the greenness index. With the roots and rhizomes remaining in the soil, the leaves dying back and regenerating topsoils, a perennial oil sunflower could be a prolific, undisturbing beauty.

The same beauties and protections come from biodiversity in perennial pastures, where mixed grasses and deep-rooted legumes cycle nutrients and build rich organic soils.

As the risks of skinhead agriculture become clearer, and consumers demand better treatment of livestock, pasture-based networks have grown up in a number of states, like Project Grass in Pennsylvania and New Jersey, encouraging intensive rotational grazing instead of confinement livestock operations, and promoting cover crops as erosion protection and green manure. Even the USDA has a few perennial-ag research projects, like the Appalachian Farming Systems Research Center in West Virginia, studying "silvopastoral" systems, pasture-based beef production, and other grazed agricultural ecosystems.

But these are dots of green in an otherwise bare-earth landscape.

Sooner or later, we'll have to cover the ground—for our own protection. There are limits to the landscapes we can scrape, exploit, and degrade without paying dearly. We are already paying, with the costs of global warming, the costs of water and air pollution, the loss of aquatic systems and fisheries, the misery of families living with CAFO/skinhead agriculture.

We spend a lot of time here trying to figure out why CAFO owners stick with a business so damaging to their animals, their fields, their neighbors. Twice yesterday, people stopped me to ask, why do you think they do it? Do they enjoy it? Is it just the money? If somebody did a psychological profile of these guys, what would it show? Psychopathology? How do they live with all the damage they do?

When you listen to CAFO owners talk, it's clear they don't agonize over harm, they don't anguish at the destruction of wetlands or woodlots or streams. In their world, human beings are, rightly, in control. They may have to scramble, out of breath sometimes, all those crises, but they're in control. They can always find answers, reasons—the fall was too wet to spray manure, we had to dump it on frozen ground, an emergency plan. They have a reason, a plan, a battery of machinery. If there's a problem, a mistake, they fix it—almost always with a more complicated plan, an add-on technology to fix the previous one. If dry manure is cumbersome and expensive to haul, well, liquefy it, pump it. And then if it's too wet and runs off and pollutes streams, well, spare no expense, construct a million-dollar liquid/solid separation system. Make some of it dry again! Pump the liquid someplace else. Build more lagoons. Never go back. Whatever was in the past, it's gone, done-for, out of date, not worth a look. Forget pastures. That's the past. Technology and machinery, that's the plan, the answer to every problem. Never Nature. Nature is no answer. Nature's the problem! Or, Nature's for vacation— somewhere else. Either way, it has nothing to do with work, it's a tangle, it doesn't draw a straight line. Nature is chaos, not order.

Where does this lead? Humans who have no faith—in fact, disbelieve—in Nature, its mesh of life-supports, its artful designs, its immeasurably complex systems, can believe only in the ordinary work of their own hands, whatever machinery they can put together. They'll work and work and see it as virtue, doing their best to plow a straight line, milk as many dollars from as many acres as possible. They'll call it God's work maybe, never Nature's work.

Dissociated from Nature, they're like the homeless. They *are* homeless. They don't recognize their own habitat, and so they destroy it. They scrape together some temporary shelter (the one $350,000 house on Dillon) and scrounge a six-figure income, often it's dirty work. But that's virtue, that's work, they don't give up, they don't stop to ask, Does this make sense? Have I lost my senses? Already cast out, like the damned, the farther they keep themselves from Nature, the easier it is to sleep.

After decades of this skinhead ag debacle, we need to look once again to Nature as agricultural model, of whole systems, not just crop-in-dirt knowledge. We need to re-cover our home ground, recover our habitat. Our agricultural policies and practices need to accept the limits of landscape and work with the laws of complex natural systems.

In Nature, the habitat we share, our only home, there are boundaries we should not cross, places we should not trespass, vegetation we should not uproot, trees we should not cut down, ground we should not strip bare.

When we come to our senses, if we come to our senses, and recognize greenery for what it is to us—lifesaving—our eyes, our vision will change. We'll see bare ground as an assault, an offense. We'll see skinhead agriculture as degrading, to Nature and to ourselves. Knowing better, we'll farm, and farm well. But we'll also be able to sit by a stream with its edges restored to forest, cool our feet in the water, and celebrate a world recovered. Not lost.